Successful Consulting For Engineers And Data Processing Professionals

Successful Consulting For Engineers And Data Processing Professionals

STEVEN P. TOMCZAK, Ph.D.

President
Steven P. Tomczak and Associates
Los Angeles

A RONALD PRESS PUBLICATION

JOHN WILEY & SONS
New York • Chichester • Brisbane • Toronto • Singapore

Copyright © 1982 by John Wiley & Sons, Inc.

All rights reserved. Published simultaneously in Canada.

Reproduction or translation of any part of this work
beyond that permitted by Section 107 or 108 of the
1976 United States Copyright Act without the permission
of the copyright owner is unlawful. Requests for
permission or further information should be addressed to
the Permissions Department, John Wiley & Sons, Inc.

This publication is designed to provide accurate and
authoritative information in regard to the subject
matter covered. It is sold with the understanding that
the publisher is not engaged in rendering legal, accounting,
or other professional service. If legal advice or other
expert assistance is required, the services of a competent
professional person should be sought. *From a Declaration
of Principles jointly adopted by a Committee of the
American Bar Association and a Committee of Publishers.*

Library of Congress Cataloging in Publication Data:

Tomczak, Steven P., 1944–
   Successful consulting for engineers and data
processing professionals.

    "A Ronald Press publication."
    Includes index.
    1. Consulting engineers.   2. Electronic data
processing consultants.   I. Title.

TA216.T65   620'.0068   82-2797
ISBN 0-471-86135-9   AACR2

Printed in the United States of America

10 9 8 7 6 5 4

To those who, win or lose, are willing to give their all for the adventures that lie in reaching for a new horizon.

Preface—A Personal Message

Today gasoline prices are sky high, mortgage money is tight, and many of us who have shifted into the 40 or 50% tax bracket are having a hard time making it. Times have changed. Engineers and data processing professionals are getting single digit raises while we have double digit inflation. I decided to do something about it and so can you. This book will help you use your experience as a technical professional to get out of this dilemma.

Many engineers and data processing professionals have asked me what it takes to go into business for oneself. Each time I try to answer this question I feel a slight sense of frustration. I try to convey how easy and profitable it can be, but my listeners don't believe me. They think that there must be some kind of mystery attached to it, and tell me, unconvincingly, that maybe some day they'll give it a try. Working beside engineers whose job was the same as mine, I have earned twice as much money and taken home three times as much after taxes. What was the difference? I was an independent, setting my own fee, and they were employees, hoping for ten percent raises.

Becoming an independent contractor is not at all difficult. If you prepare carefully and really understand what you're doing before you do it, you will be able to make an easy transition into a much more lucrative and high-powered lifestyle. I made some really serious mistakes when I started out. In retrospect, some of them appear humorous, but if you make them, you won't find them funny at all. And this is my purpose for writing this book—to help you avoid these mistakes in the preparation stages and to make this transition as smooth as possible. This book will guide you through the crucial initial stages of setting up your practice, and will help you with the fundamentals of operating successfully.

When I started consulting I had been working as an engineer for six years. I got tired of being an employee and became something of a clock-watcher. When I was an employee I simply didn't have my heart in it. My employers were fair and treated me well but I still looked forward to Fridays. T.G.I.F.—Thank God It's Friday! Whenever I hear that phrase I'm reminded of the days when I was not leading the life I wanted. I didn't become a professional to wait for Fridays so I could start living. I believe that everyone has a responsibility to himself to live the lifestyle he wants to live and to create an enjoyable work environment for himself. Even if I earned less, I would still want to be a consultant.

I'd like to tell you a bit about my background. Most authors don't say very much about themselves, and I believe this is a mistake. An important part of understanding any material is knowing something about the author.

I received my education at the University of Connecticut. I obtained my doctorate in physics in 1971, and after teaching for a short while, I went to work in industry. My first job was for Honeywell in Lexington, Massachusetts, where I gained engineering and management experience. After a few years, I took a position with M.I.T.'s Lincoln Laboratory. It was a fine position and I worked with many talented people. However, though I couldn't identify why, I began to realize that something was wrong. I didn't want to believe that all my schooling had led me to a job that was just mildly interesting and would make me look forward to Friday.

There were obvious financial considerations. Inflation and taxes were taking most of my income away from me, and I was not expanding my financial base. My survival was very dependent on the company for which I worked. My education had qualified me for a job but it apparently had not guaranteed my financial survival in any larger sense.

My dissatisfaction led me to take a leave from my position at M.I.T. I decided to do some traveling while I considered what the next phase of my life should be. My vacation was coming to a close, and despite the fact that I was running out of money, I decided that under no circumstances did I want to return to work as an employee.

I didn't know the basics of business, and I was scared to try to go it on my own. I didn't even know anything about becoming an independent contractor or consultant. Nevertheless, I set out to find a contract. I wrote a résumé, started knocking on doors, calling people, and in a rather frantic way did anything I could to try to find a lead. I had no idea of how to promote myself, how to present myself, or even where to look for work. I had an abhorrence of going back to work as an employee and a burning desire to make it on my own. Although I was short on knowledge, I was quite long on persistence. My persistence won out.

I was finally hired to solve a technical problem for a company that was started by a few of my former associates at Honeywell. My joy when I received that first purchase order was beyond words. The contract was for only two weeks, but the money was secondary. This first contract put me in business for myself, and somehow I felt that I had beaten the system.

At the beginning of that first consulting job, a friend asked me about the contract I had agreed to, and I told him that I wasn't sure whether I could solve the problem. One of the attitudes I took at the time was that the worst that could happen was that I would look at the problem in depth, and if after a few days I found that I couldn't solve the problem, I would just give the money back to the client and tell him I couldn't do it. If I could, I had

Preface

a contract. If I couldn't, the most that I would lose would be a couple of days of work. As it turned out, I did solve the problem and the client was very happy.

My second contract was for four months. I then began to learn how to promote myself properly. After that, each contract became easier to obtain.

If you're looking for a free lunch I don't recommend consulting. But if you're a maverick with a gleam in your eye and the desire to have every day be a Saturday, being your own boss may be for you. Sometimes I must work long hours running my own business, but it's a lot easier than working for someone else—and I no longer watch the clock. Sometimes there is hard work involved, but there is also a great deal of satisfaction in doing it your own way. If you decide to give consulting a try and don't like it, the forty-hour work week and the twenty-year gold watch will be there to return to. However, you may find that being on your own is exactly right for you and, if that's the case, you'll never have to say T.G.I.F. again.

My choice of the subject matter for this book was based on the information I needed when I was starting out. My objective is to save you time, money, and frustration that come from having to dig up most of the material on your own.

Since 1979, when the first version of this book was written, I have interviewed and surveyed over one hundred consultants and would-be consultants. The data I have gained through these interviews served as a foundation for this book. These chapters are an attempt to competently handle every major question commonly asked by a new consultant.

The most important of these involve promotion—how to get busy and keep busy—how to find prospects and turn them into clients. Also handled are the various technical and business elements of successful consulting: taxes, contracts, presentations, negotiations, etc.

Even though it is not difficult to be a consultant, you should still know the fundamentals of running your own practice. Avoiding some of the mistakes that I made along the way will not only save you money, but will make your transition to a new lifestyle an enjoyable experience.

STEVEN TOMCZAK

Los Angeles, California
March 1982

Acknowledgments

This book, like any other, is the product of many people and experiences. To be sure, I am not a self-made man. My success is a result of the love and assistance from many people who willingly reached out their hands when I was in need.

I wish especially to thank four very close friends: Peter Tomczak, my Dad, who always encouraged me to follow the path of my choosing; Marylyn Tomczak, my Mom, who placed in me the seed of reaching for more; Kristin, my sister, who has been a constant reminder that love is what it's all about; and my sister, Janice, who was always near me despite distance or those things in life that can momentarily take us from those we love.

The idea for this book was born out of conversations with Logan Biggs, of Newport Beach, California. Logan is a friend who nurtured my dreams with enthusiasm and, with his care and understanding, acted as a catalyst for my ideas. Special mention and thanks also go to my friend and colleague, Craig Shields, who reviewed and edited the entire manuscript. His suggestions and enthusiasm were responsible for making the project an enjoyable one.

To others, who have left their mark on these pages, I extend my appreciation—Ruth Terry Trolinger, Steven DuVall, George Duggan, Susie Gorski, Pat Burger, and Mark Jones.

Above all I would like to acknowledge and thank L. Ron Hubbard for providing me with the guidance and wisdom that prevented me from making the mistakes that are so common to the budding entrepreneur. I am delighted to see that Mr. Hubbard's insights into human behavior and business management are being recognized as the superlative achievements they are by leaders in so many disciplines.

S.P.T.

Contents

1	Consulting, a Viable Alternative	1
2	Promotion Pieces for Your Practice	17
3	Promoting Your Practice	29
4	A Promotion Program for Your Practice	45
5	Subcontracting to Job Shops and Agencies	59
6	Turning Prospects into Clients	67
7	Giving Effective Presentations	75
8	Determining and Structuring Your Fees	87
9	Negotiation	103
10	Contract Basics for the Consultant	115
11	Understanding Your Rights and Obligations	131
12	Delivering Your Service	143
13	Tax Savings and Financial Planning	153
14	Should You Form a Sole Proprietorship, Partnership, or Corporation?	177
15	Start-Up Projects to Do in Your Spare Time	199
16	Further Help	221
	Appendices: Sample Contracts	229
	Index	335

1
Consulting, a Viable Alternative

I'm an independent engineering consultant, and I've written this book to help you gain a better understanding of information necessary to start your own consulting practice. You will find that this book will help you make the decision that is right for you. Perhaps consulting is not for you; in any case it's important that you make the correct decision based on the proper data. However, you might find that consulting is the game you want to play. If that is the case, get ready for some fun, for there are few games greater than the game of creating your own independence.

This first chapter answers a few dozen of the most commonly asked questions about becoming a consultant. It is written to help you better understand the generalities of consulting and independent contracting.

WHAT IS A TECHNICAL CONSULTANT?

A consultant is anyone who is professionally and legally qualified[1] to provide advice or services to clients in areas of his identified expertise. He's usually paid proportionately to both his abilities and the demand for his specialty. (This demand, though, as we shall see, is more related to how well he promotes his expertise than the "demand in the marketplace.")

There are millions of technical specialists in the United States. Between five and ten percent of these specialists are consultants or independent contractors on a full or part time basis. However, because consulting and independent contracting are so lucrative, this number is growing rapidly. This rapid growth is accelerated by tax and inflation pressures.

If a person is a specialist in a technical field, he can provide his clients with an objective and impartial point of view with regard to technical projects and technically-related management problems. He is sometimes brought into a company to deal with upper level management or to take on

---

[1]Legal Qualifications are discussed in Chapter Fifteen.

responsibilities that form the bridge between the administrative and the technical sides of a project.

There are so many different types of consultants that it's difficult to make generalizations that apply to all areas. In addition, there are also many different types of clients and thus, to describe consulting from a general point of view is of limited value.

The reasons why a consultant is used by a client are many. Often the consultant is brought in to solve a particular problem that the client does not know how to handle. When he's brought in for this reason, the problem might be solved in a day, or it could lead to a project that could extend into a month or even a year.

Some companies use consultants to minimize fluctuations in hiring or to relieve staff members during overloads. In the past, many highly specialized companies have gained reputations for treating their specialists like cattle, herding them in and out as they win and lose contracts. In order to keep competent help, these companies now have recognized that stable hiring policies are needed. Thus, companies are willing to pay the higher prices for consultants to relieve the strains that are created during peak loads.

Consultants are also hired by clients to run entire projects. These projects can be of any type and can vary in size and complexity. Some consultants work both as technical specialists and as program managers on technical projects.

Although the consultant's expertise is often publicized as being technical in nature, his advice in financial matters is constantly sought. For example, a consultant is often brought into a project in its developmental stages to generate cost estimates in order to assist in feasibility decisions and bid preparation.

It is believed by many that the primary requirement for successful consulting is technical expertise. While this might be a necessary condition for his success, it is certainly not sufficient. The presence of the consultant should bring more predictability and certainty to a client's decisions. In order for a consultant to be effective in this regard, he must have a high level of professional integrity. Often, a client regards the consultant's professional integrity above his technical ability. Another important factor which the successful consultant must demonstrate is his ability to work efficiently and effectively with the client's staff. The general rule is that *the successful consultant must be able to identify the critical objectives of his client and direct his attention and energies into activities that will assist the client in reaching his objectives.*

Once a good working relationship has been established with a client, the consultant is often kept on a retainer or is employed periodically by the client. In the long run, the retainer method often saves him a great deal of time and money. To a client, the greatest benefit of a consultant is often obtained after the consultant gets to know the client's staff and operation. At that point, the consultant can begin to deliver his most effective advice and service.

WHAT IS AN INDEPENDENT CONTRACTOR?

The word "consultant" implies a person with a specialty who plays the role of an advisor to a firm. The firm decides whether they should take the consultant's advice. Often the consultant is not responsible to see that his advice is carried out. In contrast to this is the contractor or contracting firm. A contractor is usually a specialist in a technical field who gives advice *and* takes responsibility to see that his part of the project is completed by himself or his own staff. There are many contracting firms of all sizes. The larger contracting firms are often seen in the construction industry where one finds examples such as: electrical installation contractors, plumbing contractors, air conditioning contractors, etc.

The independent contractor is often a technical specialist such as an engineer or a programmer who is self-employed and takes on temporary project assignments for a company. He often does the majority of his work at his client's location and arrives for work every day similar to the client's employees. He often provides temporary help, like an engineering job shopper, with the major difference being that he is self-employed. Many job shops are now hiring independent contractors and then subcontracting them out to their clients.

Examples of independent contractors are: circuit designers, draftsmen, programmers, thermal design engineers, and digital designers. The keynote that identifies the independent contractor is his self-employed status and his willingness to take on temporary assignments. The income of the independent contractor is usually twice or three times that of his salaried peers.

Because independent contracting has become so popular lately, many independents are taking work assignments and are completing them in their homes. An example of this is the larger number of computer programmers who have their own computers and terminals in their home or office. In cases such as this they often take on a contract on a fixed fee or hourly basis and do all or most of the work at their office or home.

There are many independent contractors who also call themselves consultants. In many disciplines the words are used interchangeably.

SHOULD YOU INVEST IN YOURSELF?

When I was an employee, I used to sit around the lunch table and talk with the fellows about the Dow-Jones Industrial Average. After this went on for a few years, I began to realize that I had met only a few people who made a significant amount of money in the stock market. After giving the situation more consideration, I decided that the best thing to invest in was something I knew a little bit about: my own specialty. Investing in myself really paid off for me, and it will for you as well.

If you're considering making some investments, I suggest investing in yourself. You have a specialty and, no doubt, if you could form an organization properly, you would find that this specialty is worth a great deal to yourself and others. Invest in yourself—you're not a bad risk at all.

In the competitive business world there is a group of very shrewd and productive businessmen in constant search of new products to promote. These men are experts in taking an unknown product that has potential and making it known to thousands or even millions of customers. Their technology is promotion, and they're always in search of a good product.

The technical specialist is often in just the opposite situation. He already has a service to offer which is represented by his experience. However, he is often lacking in the know-how of getting his expertise known to a sufficient number of people who could become clients. If you're a specialist, you should recognize that you have the potential for delivering a valuable service. Consider yourself fortunate that you have a product that is something of value.

There are many non-technical businessmen who would give their right arm to have a product that is as solid as one represented by your expertise. The important thing to recognize is that though you might be ignorant of how to run a business and how to promote yourself, you *do* have something with which you can start a business. Take it easy and start accumulating the necessary facts to make your own business come into being. There's lots to learn. By being patient and collecting enough of the proper data you can become successful.

If you learn the necessary ins and outs of the business world, the only thing that could hold you back is having a specialty that is not in demand. You'll be able to find out whether you are in demand if you go along with the promotion program that is described in this book. You've invested a great deal of time and energy to develop your expertise. You have a prized possession. If you want to make it worth even more, recognize the fact that you're going to have to learn how to play the game of being a businessman. If you're going to operate a business, you're in the game of surviving by making a profit. It's a new viewpoint, and with a little practice you can become as good as the best of them.

After you've made it through the first few years and have developed a solid foundation of four or five reliable clients, you will reach a point where you can expand and form a larger, more stable operation. If you can make it through your first year or two successfully, the chances of failure are considerably reduced. This is why a solid preparation is so important. By working full-time and setting up your consulting company on a part-time basis, you can save yourself thousands of dollars and increase your chances for surviving through the first and toughest year. Take a conservative approach by starting out slowly and carefully. This might not be as glamorous as other approaches, but it is certainly a path that will lead you to success.

HOW MUCH CAN YOU EXPECT TO MAKE?

A technical consultant or independent contractor can generally expect to gross two to three times what he would gross as an employee. His take-home pay after taxes will be even greater because of his tax deductions. The tax advantages are discussed in Chapter Twelve.

A typical programmer earning twelve to thirteen dollars per hour can expect to make twenty-four to thirty dollars per hour if he promotes himself properly. Likewise, an experienced engineer making twenty dollars per hour can shift to the forty to fifty dollar an hour bracket.

Another advantage to being self-employed is that the consultant is often included in partnership arrangements and in money-making activities that are often not available to an employee. An example would be the case in which a client tells you he needs more manpower. In this case the consultant might bring an associate on the job and take a percentage of the associate's income as a subcontracting finder's fee.

Many consultants and contractors charge time and a half for overtime. If you are in the kind of work where overtime is common, you will find that consulting can be quite lucrative. A small four week contract with twenty hours of overtime per week will bring $14,000 at a base rate of $50 per hour.

WILL CONSULTING BE SIMILAR TO YOUR PRESENT WORK?

Some believe that in order to be a successful consultant, you must have some kind of super specialty or possess some kind of magic. Nothing could be further from the truth. The most surprising thing to me about technical consulting is the fact that many of the activities which I go through in my consulting work are very similar to the work I used to do as an employee. All I'm doing is delivering technical services and advice to clients. A large portion of my time is spent solving technical problems and then communicating effectively with my clients. Doesn't that sound like what you are already doing on your job?

Of course, there are many new tasks and activities with which I have to be familiar. I must keep books and I had to learn a little bit about contracts and liability insurance. But there is no magic associated with this. After you pick it up, you know what to do with it, and then it just becomes another working tool. Once you know a few basics you can deliver your services in a similar fashion to an employee, with the major difference showing up in your savings account.

If there is any secret at all to this business, it is that there is nothing to fear; there are no boogymen out there—really. What's required is that you exercise common sense and recognize that you might not know a lot about the business world. In fact, recognizing that you don't know much about the business world can be an effective asset. The business world is filled with people who believe that they really know what they're doing and they are not hunting for new techniques for making their business succeed. One of your assets is the fact that you know you are new to this game and that you're willing to go through the necessary activities to find out what you have to do to operate a successful business.

HOW IS A CONSULTANT PAID?

The standard methods of payment are: (1) fixed fee, (2) cost plus fixed fee, (3) per diem, (4) straight time, (5) retainer, and (6) fixed percentage of the cost of the project. These methods are discussed in detail in Chapter Seven.

ISN'T THERE A LOT OF COMPETITION?

Since consulting usually involves highly specialized work, a consultant does not usually compete with other consultants in the area of fees for doing a particular job. This does not mean that the consultant refrains from competition; he certainly competes with other consultants, but the competition is more in the area of delivering quality services. Consultants are hired because of their integrity, technical expertise, and administrative abilities. When formally bidding for a job, these are the factors which have to be stressed. If a consultant bids to do a job and is chosen strictly for "lowest price" reasons, he's probably dealing with a client who is unfamiliar with working with a professional consultant.

WHAT DOES A CLIENT LOOK FOR IN A CONSULTANT?

There are many factors which come into choosing a consultant. We will treat them in order of importance.

Delivering What You Promise

A client is always looking for a consultant he knows will be honest with him and who will be worthy of his trust. He wants to have confidence in everyone with whom he works so he can have predictability in his programs. However, as it is difficult to determine a person's honesty or confidence level immediately, the only thing that he has to go on is your past record of what you've promised and what you've delivered. Having letters of recommendation available can be beneficial in demonstrating to him that you mean business and that you can complete the job.

Technical Competence

The potential client will be concerned with your technical competence to do a particular job. He is also looking for the extra competence, intelligence and persistence required when things are not going well on the program or when unexpected barriers arise. If he hasn't dealt with you before, he has to rely on the technical description of your past experience. Again, as I state so often through this book, you must have references and letters of recommendation to show people that you really can do the job.

Management Abilities

In some cases a client is looking for a strong leader with the ability to get along with his employees. Very few technical consultants work in a vacuum. Most often they have a strong interaction with the client's employees and the employees of other companies. The consultant is often expected to direct and coordinate others, as well as guide the program technically in order to get the job done correctly. A consultant with excellent technical credentials has less value to a project if he cannot get along with and lead other people. The consultant must be highly aware of how his role fits in with the roles of other employees. He must also have the confidence and practicality to ask the proper authorities how he should fit into the company structure to be most effective.

DOES A CLIENT OBJECT TO A CONSULTANT'S FEE?

Note that in these three important points above we haven't mentioned the fees that the consultant charges his client. The reason for this is that, although fees are important, they are not a prime consideration when a client decides on a consultant. Most consulting prices fall within a reasonable range, and clients who have dealt with consultants are often well prepared to pay quality prices for quality and delivery. This does not mean, of course, that the client is willing to pay extremely high fees for a consultant just because he's ethical and competent. *However, you should charge what you believe you are worth and not accept anything less*. If you deliver a good product and you deliver what you promise, you're worth considerably more than the average employee or consultant. On the average, the talent available to most clients is pretty low in quality, and if your emphasis is on delivering what you have promised, then you really have something to offer that's unique. So, don't be afraid to ask for a healthy fee and make a profit that is in proportion to your worth. Successful businesses are based on delivering quality products and obtaining excellent profits in return.

My experience has been that many of the clients for whom I have worked have considered becoming consultants themselves. They are not only not surprised with your fees, they often wish they were in your shoes. You are apt to generate respect from them rather than scorn. Remember, they too are in the same financial trap that limits most employees.

WILL CONSULTING HELP YOU HANDLE INFLATION?

Yes. As inflation erodes the dollar, your rates can be increased proportionately. When you are in an employee capacity your pay raises are determined by the financial condition of the company and by your supervisor. When you're in business for yourself, all you have to do is raise your rates. If raising your rates starts to slow you down in getting new contracts, you just boost your promotion activities so you will have more clients from which to choose. One of the greatest assets of being in business for yourself is that you determine your own fees.

The biggest problem that I have observed in helping new consultants get started is that they underestimate what they can get for their services. When you start out you should set your fees so that you are at least doubling your present salaried income.

WILL YOU SAVE ON TAXES?

Yes. In the materials that follow, you will see that you can save plenty on taxes. There are many tax savings that you'll be able to make because you are self-employed. You will be in a higher tax bracket and you will need some professional financial assistance. However, these are a new set of problems most people are delighted to take on.

I have a lot of business tax write-offs and Uncle Sam doesn't gouge me like he did when I was an employee. It's true that it's a little difficult sometimes to set up my finances and invest my money in such ways that Uncle Sam doesn't really sock it to me, but the fact of the matter is that I now have more control over my income. When I receive my income I've got it, and then I have to figure out ways to set my finances up so Uncle Sam doesn't get it all. This is a lot different than having the money taken out of my salary and then trying to get the money back from the government.

I've overheard people say that they were really happy this year because they were getting $400 or $500 back from the IRS. They're acting as though the government were giving them a break because they were getting $500 back on their tax return. It's almost as though they'd forgotten that the government had taken out $7000 or $8000 from their paychecks, and all they were looking at was the $500 return. That $7000 or $8000 that the government had taken out over the course of a year was gone forever and there was no chance of being able to get it back. I like having control over my money. I don't like the idea of the government taking most of it away from me. And by God, I do everything I legally can to make sure that the government doesn't get it all.

HOW DO CONSULTANTS GET STARTED?

Often the consultant in private practice has worked as an employee for many years and has gained considerable experience and expertise in one or two specialty areas. He has contributed to many projects and has gained the reputation of being able to get a job done. Sometimes he starts consulting on a part-time basis and gets the feel of consulting slowly while he has some extra income coming in from his regular position.

It rarely happens that a successful consultant just decides to start his own business, hangs out his shingle, then waits for clients to show up at his door. Probably the most important contributing element to successful consulting is promotion. Promote when you're small; promote when you're big. Promote when you have no clients; promote when you're busy. The fact that many consultants are so specialized that they are not even listed in the Yellow Pages of the phone book does not mean that they're not known. In fact, it is probably an indication that they're so well known that they don't have to advertise very much.

Just because a minority are not actively promoting, don't be led to believe that advertising and promotion are unimportant. Most successful consultants are constantly promoting. If you asked a consultant about promotion, he might not even be familiar with the word. But, because he participates in technical societies, publishes papers and is in contact with a large number of potential clients, he's often forming a very solid promotion foundation.

HOW MUCH MONEY WILL IT TAKE TO GET STARTED?

There are some people who have started working as an independent contractor or consultant without spending any money. They simply called somebody up on the phone and asked them if they needed help. The new client said yes, and they went out and did the job, billed the customer, and that's all it took to get started. There's probably not a business practice that exists that is cheaper to start than consulting. Likewise, there's probably no business easier to get out of than working as a consultant. There are a lot of businesses that get started on a kitchen table. If you really want to get down to basics, you don't even need a kitchen table to get started. All you need is someone to offer you a contract.

However, being practical, it will take you a little more than that to get started. In the first place, I suggest that you get started in your home or in a small office. This is going to require a desk, some filing cabinets, a phone, an answering machine, and some business cards. But even these things can all come later if they have to.

I definitely recommend starting off on a part-time basis, and I do not recommend spending a lot of money to do so. When you're getting started, something you're going to have to remember is to keep your overhead as low as possible. The thing you don't want is a high overhead for things like an office and rental furniture. In the beginning you should keep your expenses as low as possible so that after you have finished your first consulting contract you won't be faced with any bills in a period where you might be looking for another job. You should never be under pressure to spend money on items that are unnecessary. If you have a fairly large apartment or a good-sized home, you can make one or two rooms into an office. Office rent is often expensive for someone just getting started, and by the time you have furnished the office, you might be spending more than is necessary. In the beginning, keep your expenses down. If you have children and you think the children are going to interfere with the business, don't worry about it. Set it up in such a way that you can either share an office with someone else, or set up a room in such a way that your children won't bother you. I can guarantee you that by starting off with a low overhead you'll be much better off and you'll have less headaches.

Once you have established a few steady clients and your income is somewhat predictable, you can move to an office and place your shingle on the door.

HOW LONG WILL IT TAKE TO GET STARTED?

I recommend to most people that they get started on a part-time basis, and that they start *now*. If you start on a part-time basis where you only have to work five or ten hours a week, then you will get your feet wet without disturbing your financial stability. This approach will also give you a good feel for consulting, and you will see if you like being in business for yourself. There are books on the market that will tell you that you have to be in business for a year or two before you can really make a success of it. That is just not true. If you know how to promote yourself, you can get started profitably on a part-time basis within a couple of months. It is simply a matter of getting your first contract and then just improving your promotion methods and expanding your client base.

HOW CAN YOU INSURE GETTING NEW CONTRACTS?

One of the fears that's often expressed by someone who's just about to go into consulting is the unknown factor about where the next contract is going to be coming from. I recently offered a consulting contract to a fellow who was working as a full-time engineer. I offered him a contract that would be for 6 months to a year, and the first question that he had with regard to this contract was, "After this contract is up, what do I do?" Well, one of the ways that you handle these fears is to promote yourself properly such that you'll always have more work than you can really handle. Another way to deal with this problem is to take longer consulting contracts. There are some people who enjoy rather short consulting contracts, and they charge much higher fees for the small amount of time that's involved. However, there are other people who will take very long consulting contracts—sometimes as long as a year. When you're taking on longer contracts you certainly don't have to deal with the question of where your next job is going to come from as often. *Your fears of where your next job is going to come from will be handled by knowing that you know how to promote yourself properly.*

If you do know how to promote yourself properly, you'll have more security than somebody who has a full-time salaried position. When you're consulting to a firm on a contract that will end at a specific time, in six months, say, you know you can start looking around for another contract a month or two before the contract ends. Contrast this with a person who is employed full time. Often he thinks that he's very secure, but then to his surprise he gets a slip in his pay envelope saying that in a week or two he's going to be laid off. Now in a situation like that he really had no time to prepare for looking for a new position. In consulting you don't have to worry about this. As long as you know that a contract is going to end in six months you'll be able to predict your future and have plenty of time to line up a new contract.

ISN'T PROMOTING YOURSELF NON-PROFESSIONAL?

If there's anything that makes me mad it's just this one question. Somewhere along the line some real dimwits have told professional people that if they promote themselves it's non-professional. This is ridiculous. If you're an accountant, lawyer, physician, engineer, or a scientist and you go into business for yourself, how do you expect to get business? You must promote yourself. You're not dirtying your hands if you promote yourself.

The thing that you must remember is that when you go into business for yourself you're first and primarily a businessman. Who ever heard of running a business without promotion? Can you imagine starting a company that was going to build computers and announcing to the world that you weren't going to advertise? The boys in the white jackets would probably be at your doorstep, and for good reason. My opinion is that it is non-professional if you don't promote yourself. If you have anyone telling you that you shouldn't advertise or promote yourself, I can guarantee you that their one concern for you is the destruction of your career and your business.

IF YOU WORK FROM YOUR HOME, WILL CLIENTS THINK YOUR BUSINESS IS EXTREMELY SMALL?

The answer to this is yes, you will give others the impression that you're small. Your answer to anybody who asks is, "Yes, I work out of my home, and I can keep my overhead nice and low. This allows me to only charge $50 an hour instead of $60 an hour."

One cannot expect a businessman to be totally unconcerned with the question of how much your service will cost him. If you can prove to him that you can do the job and you can charge a lower fee because of your low overhead, he'll be delighted that you're just a small-timer working out of your home.

Another factor to consider is that many people enjoy working out of their homes. There are no trips to the office. If you think of something in the evening that you'd like to do for your business, all you have to do is walk into the next room and get it done. It's fun to work out of one's home. And it can also be very pleasant for your clients.

I often have a client come to my home for a visit or for dinner. It's always very relaxed and informal. We often go into the living room and chat for a while before having dinner. I've never met a client who didn't enjoy the informal atmosphere.

A misunderstanding exists in many small businesses that are just starting out. A lot of people think that if the clients get wind of the fact that you are small and you haven't been around for a long time, that they won't take you on. This just isn't true. You don't have to promote yourself with stability symbols like the Prudential Insurance Company with its Rock of Gibralter. All you have to do is present yourself as a person who is enthusiastic about doing a good job, and as someone who has the ability to deliver what is needed and wanted. I have never met a client who didn't want to give me work because I was operating on a small-time basis.

HOW MUCH EXPERIENCE DO YOU NEED?

There is no pat answer to this question. The answer is really different for each individual person because there are so many factors. First let me say a couple of things up front that I think are very important. You're not going to be a successful contractor or consultant unless you really have something to offer somebody. There are some people who claim that you can make a hundred thousand dollars a year peddling your advice no matter who you are. Personally I don't believe it. Working as an independent contractor or consultant isn't just peddling your advice. You're going out and solving problems for people. You're finding what they need and want and you're delivering it. This requires that you have some expertise in a particular area. Be it finance, law, engineering, or science, you have to have some area of expertise where you can fill a real need.

If you are going to be working in industry, the chances are that you already have a typical education in your specialty and probably a minimum of four or five years of experience. You can act as an independent contractor in computer programming with only a few years of experience, provided you know how to operate a particular system that's in demand. After all, if a company's just spent a quarter of a million dollars on a mini-computer and they're losing money because it's not programmed in time, you can bet your boots that you'll be hired to help out on such a problem, even if you're rate is high and your experience is low. In this case the demand factor is taking care of you.

However, there are many of us who work in areas where the demand is not as critical as it is in the computer industry. In such instances you still want to know how much experience is required.

One of the first questions that you should ask yourself with regard to experience is how well do you perform on the job now. What's your work history? How many years of experience do you already have? Let's say that you're an engineer or a programmer and you already have ten years of experience. What's your record of achievement? First of all, take a look at the other engineers around you and ask yourself the questions: On the average, do you do a better job than they do? Are you brighter than the rest of them? Look to see if you do a more conscientious job than most of your peers. Again, I'm not asking you if you make more money than they do. The critical question is, can you find what's needed and wanted from your employer and deliver it in a better, more efficient way than other employees? This is a very important question because I don't think it's the years of experience that you need. It's the ability to see what's needed and wanted and deliver just that. If you can deliver what is really needed and wanted by your client then you'll have very little competition. I don't care how many people are in competition for the same position. Most people do not deliver exactly what is needed and wanted; they are mainly concerned with making money. Managers and prospective clients know this.

A consultant eventually becomes successful based on his reputation. If you commit yourself now to technical excellence combined with professional integrity, you'll be setting a true foundation for your business. Honesty with one's self and with one's client goes a long way. If you maintain high technical excellence and professional integrity, you'll be able to compete with a broad spectrum of individuals and companies which offer similar services. Honesty is a rare commodity in business today and the wise client knows how to recognize, appreciate and utilize it.

DO YOU NEED A STRONG REPUTATION?

You do not need a strong reputation to become successful. Although a reputation helps, it is not necessary. If you've written a lot of research documents, if your name has been in the paper a lot, if you're the author of a textbook, if you've written many journal articles over the years—sure that helps. But let's look at this in a little more depth. Why does it help? It helps because being an author or being in the limelight is, in effect, promotion. It's simply that your name and your abilities are being promoted to a large number of people.

Now, whether or not you wrote those articles or textbooks for promoting your business is not the issue here. The issue is that a reputation is something that you create. The point that I'm trying to make is that if you don't have a strong reputation in your field, the way you handle that is that you just promote yourself as a person trying to do a good job, and I guarantee you that if you follow this simple rule you're going to get all the work you need. There are even situations in which not having a reputation actually helps. Some clients will want to hire somebody who will do a competent job and not be charged at the highest rates. Somebody with the biggest reputation in the field might be charging $100 an hour. Somebody else who can just do a competent job might only charge $50 an hour, and there are plenty of clients who would much rather spend $50 an hour for somebody who is competent instead of $100 an hour for Joe Famous. This business about needing a super reputation is just a lot of hogwash.

The reason for this misunderstanding about needing a reputation is that the people who often have big reputations are the ones who are in the limelight, and they're the ones you often see. However, there are plenty of other guys out there making very good money who don't have super reputations. Just because they don't have visibility doesn't mean that they are not there. So, again, what we've come down to is that promotion is what is necessary, not a reputation, a big fancy office, or any other nicety.

WILL YOU BE BURNING YOUR BRIDGES?

A common fear and misconception that people have before they go into business for themselves is the feeling that they're giving up a career or that they're giving up a job with their company. It's not true that to go into business for yourself you have to leave your job. But even if you do leave your present position, remember that if you leave properly, you will always have the option to go back to your job or to go back to a similar full-time position. If you "up and quit" without any notice to your employer, or if you upset your employer by leaving a project undone, then you will have burned your bridges, and you won't be able to return from whence you came. However, if you leave your position in a responsible fashion, not only will you have not burned your bridges, but you might also have paved the way for coming back into a much better and higher position if you decide that being in business for yourself is not for you.

It's a complete misconception to think that going into business for yourself will place a negative mark on your career if you decide to go back to work as a full-time employee. I've been associated with companies that have rehired people who have been in business for themselves. In every case, the company that was hiring the individual thought more of the individual because that person had the willingness to try to go out and start his own firm. People respect you for trying to make it on your own. You'll even find that some companies will hire you knowing full well that within a couple of years you might be back out on the road trying to go into business for yourself again. Companies know that people who are in business for themselves are usually more aggressive and more ambitious, and can contribute more than the average employee.

HOW ABOUT YOUR JOB SECURITY?

Firms are constantly hiring young engineers and replacing the older ones. It's less expensive and it seems to be a characteristic of industry. Many say that this creates a job insecurity. I'm not going to get into any debates about job security, though I'm pretty opinionated about the whole area. As far as I'm concerned, the only job security that you will ever have is created by you. Job security is not something that is obtained by working from 8 to 5 for a firm which you feel is giving you security. I believe security is gained by creating a valuable product or service and delivering it, by knowing how to survive by providing a valuable service. Again, if you're looking for more security, I believe it's similar to investment. As you and your skills are the source of your security, investing in yourself is a good start for creating more security for your future.

The greatest insecurity comes from working for an employer in the same department for many years. In a case like this your abilities get very little outside exposure. If tough times come along, the person who has been with the same company will have very few contacts to assist him. Being certain you can make it on your own is security no one can match.

This whole concept of job security is a real weird one. Job security is not obtained by working for a company for 30 years and receiving a gold pin. My opinion is that *job security can be gained by knowing many people and having lots of contacts*. If tough times come and, if one of your clients can't take you on, then somebody else you know will. The only thing that's going to do that for you is having many contacts. A large number of contacts comes from heavy promotion and working for a large number of different clients.

WILL YOU HAVE TIME FOR OTHER BUSINESS OPPORTUNITIES?

If you want to create time for other business opportunities you can take about 25% of your gross income and put it into a savings account which can be used as a buffer for periods when you're not working. For example, if you're making $1600 a week as a consultant, you might want to take $400 a week and put it aside into an account and then use this money when you're not consulting. What this would mean is that for every three months that you consulted, you'd be able to take off one month in between consulting jobs.

You might find that taking this time is actually more profitable than working. The reason for this is that as a consultant, you will be introduced to many business opportunities you would never meet up with as an employee. For example, other engineers and businessmen will often bring new product ideas to you to develop—maybe a new electronic device or a new computer program. You will be provided with these opportunities and you might want to put some time aside to work on them. During these times your income might not be high because you won't be getting an hourly fee for consulting. However, in the long run, the fees that you might earn for producing a new product will be substantially more than you ever dreamed about making as an independent contractor. In short, by creating more time off by consulting, you will have the opportunity to venture into new business opportunities. Consulting is a road to new opportunities.

WHAT IS THE LIFE-STYLE LIKE?

Another thing I would like you to look at is: are you satisfied with the work that you are doing? Do you actually like working in your present environment or would you like to be on your own? One of the things that is very important to me is that I like the idea of making a good income when I feel like working. I like being able to take time off to be with my family and do other things when I choose to. I have the freedom to work when I want to work and then, when I want to do something else, I do it. I never liked the idea of having to wait for my two or three weeks of vacation to do the things that I wanted to do. The way I see it, there should be 52 weeks in a year to do what you want to do. I'm not making the claim that every moment in my life I'm doing exactly what I want. But I'm certainly doing more with my life in terms of what I want than when I was an employee. When I was an employee I had a tendency to look at the clock and do what I wanted on nights and weekends. Now, I'm in business for myself. If I want to take time off or if I don't want to take on a contract, I don't have to. My income is sufficiently high such that in four to five months I can earn enough for the rest of the year. I not only have more money, but I now have more time to spend on the projects of my choice.

2
Promotion Pieces for Your Practice

These next five chapters help the new consultant handle what is probably the biggest fear associated with going into business for oneself — promotion. "How will I stay busy?" "Where will my next contract come from?" The answers to these questions lie only in a true understanding of successful promotion. That is, these questions boil down to only one: "How many people know about my firm and want my services?"

Part of any successful promotion program is creating the promotion pieces that will be sent to prospective clients. For many applications a résumé and a cover letter are sufficient. However, a résumé is often supplemented with other promotion materials that you can easily put together in your spare time. This chapter highlights the most important promotion pieces that will assist you in finding prospects and turning them into clients.

THE RÉSUMÉ

For most technical professionals who become consultants or independent contractors the résumé is one of the key promotion pieces; it is the promotion piece that is used by most individuals in private practice. It is used more often and more effectively than expensive brochures. The reason for the popularity of the résumé over other types of promotion pieces is that it does the job of describing your abilities most effectively, and it is the promotion piece that managers are most accustomed to reading.

The résumé that is used by a consultant or independent contractor is somewhat different than the résumé used by someone looking for a salaried position. Here are some suggestions on what should and should not appear in the résumé.

You must include:

1. Your name, address, and telephone number. Make sure the number

you give will always have someone present to answer any calls. If you do not have a full-time secretary, make sure your phone is connected with an answering service or to an answering machine.

2. Brief description of your work history: The length of the description of your work history should be governed by the fact that ==the length of the résumé should be no more than two pages==.
3. Professional licenses: If you are a professional engineer, include the subject area and the Professional Engineering Registration Number.
4. Security clearance: This is not necessary if you do not work in facilities that require a clearance.

You may or may not include:

1. References to publications: This can be put to a separate sheet. If the publications are of a highly specialized nature, their subject matter might not be understood by the manager reading your résumé.
2. Membership in professional organizations.
3. Academic honors.
4. Your objectives with regard to the types of assignments you prefer.
5. Brief personal history.
6. Military service record.
7. Key words at the top of résumé summarizing your abilities.
8. Statement of health.

You must never include:

1. Reasons for leaving past employers.
2. Past salaries.
3. Current consulting rates: Leave this to a separate sheet.
4. Highly technical terms that clients won't understand.
5. Acronyms for projects with which clients are unfamiliar.
6. A photograph of yourself.
7. Names, addresses, and telephone numbers of references. Leave this to a separate sheet.
8. Names of spouse and children.
9. Marital status, number of children.
10. Your age, height, weight.
11. Hobby information.

The Résumé's Purpose

==The purpose of the résumé is simply to help you obtain an interview.== It should be long enough to tell them that you are most likely the person they need, and it should be short enough so they want to know more about you. ==Remember that your résumé is a personal advertisement, and like any==

successful ad it should be attractive. It should be typed neatly or typeset. It should create interest in you. This is done by emphasizing that you are an achiever and that you have a history of delivering an excellent service.

Make the résumé easy to read. In the sample resumé that I have drawn up, note that I have included some buzz words at the top. If a large number of people are to read your résumé, these buzz words will catch their eye and they will want to continue reading.

When describing your work history you should start with your present or most recent position and work backwards. It's not always necessary to put the exact dates of your employment or past contracts. Details such as this can be given to a client if he needs them.

In describing your work experience you should avoid pronouns such as I, he, or she. For example, don't write:

"I designed a digital logic circuit."

but, instead,

"Designed digital logic circuit."

Note: In describing yourself in a more personal paragraph near the top of the résumé you may use pronouns. As an example, see the sample résumé in this chapter.

It is not necessary to go into great detail in describing your education. Unless you have just finished college, it is your experience that must be stressed. If you don't have a lot of experience, then describing your education is acceptable.

Letters to Accompany Your Résumé

You should include a brief cover letter or a brochure with your résumé. The purpose of the cover letter is to assist you in delivering your message directly to a specific person. It should not be long. Get right to the two main points: That you can assist the prospective client in reaching his objectives, and that you are available, or will be available, on a certain date.

If you know someone who is acquainted with the prospective client that will be receiving your résumé, then by all means include his name. The first line could read: "I was speaking with Pete Simpson last Friday and he recommended that I drop you a line . . . ." Referrals are very important, and you will get a much higher response from clients if you use them.

OTHER PROMOTIONAL NECESSITIES

Letters of Recommendation

An important promotion piece which you can use while interacting with prospects is a notebook which is filled with letters of recommendation.

DR. STEVEN TOMCZAK, 6969 West Main St. Los Angeles, CA 99999 (111) 111-1111

SYSTEMS ANALYST AND DESIGN ENGINEER

SYSTEMS DESIGN AND SIMULATION	SYSTEM ERROR ANALYSIS/BUDGETING
ALGORITH DEVELOPMENT	SENSOR AND SYSTEM SPECIFICATIONS
SYSTEM PERFORMANCE ANALYSIS	TEST PLANNING
MATHEMATICAL MODELING	TEST DATA ANALYSIS
SIGNAL/NOISE ANALYSIS	STATISTICS AND STOCHASTIC PROCESSES
SCIENTIFIC PROGRAMMING	SOFTWARE DEVELOPMENT

Dr. Tomczak is an engineering analyst with over 12 years of experience in systems engineering and design. His strong background in physics, mathematics, and engineering combined with hardware and software experience, provide a foundation for getting practical results rapidly. He is recognized for his rapid adaptability to a wide range of engineering problems. Much of his experience has been in electronics and electro-optics, and he has also worked in many other areas, such as thermal analysis, radar, reentry dynamics, communications, and signal processing. He is always interested in challenging assignments and is quite comfortable taking on problems in areas that are foreign to other engineers. He can work independently, as a member of an engineering team, or as a team leader. Dr. Tomczak works as an independent contractor and is available to work on site on a full-time basis. He has a SECRET clearance.

EDUCATION

UNIVERSITY OF CONNECTICUT: B.A., M.S., Ph.D. Physics (Electromagnetics); 1966, 1968, 1971.

EXPERIENCE

TAYLOR AND ASSOCIATES, Los Angeles, CA (1980). Assistance with a space shuttle contract to determine the optimum use of various payload configurations.

HUGHES AIRCRAFT—SPACE AND COMMUNICATIONS GROUP, Los Angeles, CA (1979). Dr. Tomczak led a team of engineers and programmers which provided the systems analysis, test procedures, and software for the prelaunch testing of four NASA weather satellite payloads. The four payloads were: an electro-optical scanning camera which produces the weather photos seen on the daily weather reports, an X-Ray sensor, and two high energy particle detectors.

AEROJET ELECTRO-SYSTEMS, Azuza, CA (1978). Analysis and design of a system that protects satellites from high energy laser radiation. Systems analysis and trade-off study to determine the feasibility of an optical satellites filter mechanism.

CARSON ALEXIOU CORP., Newport Beach, CA (1976–78). System design of an optical filter for space probes and sensors. Polarization system design. Analysis of the interaction of electromagnetic radiation with high-frequency acoustic waves.

MASSACHUSETTS INSTITUTE OF TECHNOLOGY—LINCOLN LABORATORY, Lexington, MA (1974–76). Conducted the design and analysis of electromagnetic sensors which included: laser radars, optical cameras, coherent and non-coherent receivers. Determined the turbulant atmospheric effects on electromagnetic wave propagation. Analysis to determine target signatures of reentry vehicles and decoys. Analysis of radar-noise sources.

HONEYWELL RADIATION CENTER, Lexington, MA (1972–74). Analysis and hardware testing of satellite sensors and modules. Design and test of semicondustor lasers. Fast Fourier Transform methods for sensor data reduction.

HAWTHORNE COLLEGE, Antrim, NH (1970–72). Associate professor of physics.

Make a list of everyone with whom you have dealt in the past. This includes past professors, past employers, past working associates, and past and present clients. Get letters of recommendation from every one of them. Write to them and tell them that you're starting a private practice and that you need letters of recommendation to help you break into the business. You will find that they will want to help you and be part of your success.

A letter of recommendation is a piece of promotion. It has always been a piece of promotion, and it will continue to be used for that purpose. The point I'm trying to make is that you must recognize the importance of *using* the letters of recommendation—if they are going to be effective. Having them in a notebook and presenting them to prospective clients is an effective technique.

If a prospective client doesn't know who you are, what do you expect him to do—take your word that you will do a good job? He's probably been burned many times, and he's no doubt cautious about working with someone new. The thing that will ease his mind is the truth, i.e., the truth about your competence. The truth does not have to come from you; it should come from your past associates who write the letters of recommendation.

Start building a portfolio of these letters of recommendation now. Just two or three letters can be very helpful. After being in business for a while, you will have many from which to choose. But for now, even two or three will be enough.

Upon completing a contract, make sure you ask your client for a letter of recommendation. Remind him of the time he hired you and you showed him your notebook of recommendation letters. He will understand, and he will be happy to help you if you have delivered a good service.

When he writes you a letter, it will also clarify in his own mind that you did a good job. If a consultant is required in the future, he'll be sure to call the person for whom he wrote a recommendation.

List of References

You should never include the names of your references in your resumé. Instead you should make a list of your references and only hand this list to prospective clients who are serious about giving you a new contract. Before releasing the names of your references you should get their permission, and after doing so, you should stay in communication with them. It is also important to insure that your references can be reached easily and quickly. This can be done by making sure you have their correct address, department number, mail station number, telephone number, etc. Remember that the people you use as references often change positions and often get new telephone extensions. Don't make your prospective client hunt down your references.

In some cases you might want to state the position that your reference holds in his place of employment. An example of a reference format might be:

Name
Position in Company
Name of Company
Department Number
Address
City, State, Zip
Mail Station
Telephone number and phone extension.
A one-sentence description of the type of work you performed.

Note that you should briefly describe the type of work you performed for your past client. This will make it easier for your prospective client to communicate to the person you gave as a reference.

If a person you gave as a reference has changed his/her name through marriage or for any other reason, be sure to modify your reference list. Likewise, if you change your name, make sure you call all the people you are using as a reference and let them know that you have done so. If you don't, your prospective client might be told by your references that they have never heard of you.

Summary of Past Contracts

You should have a brief summary of past contracts available, not to be mailed out in volume like your resumé, but to be used when meeting with a potential client who is seriously interested in hiring you. There are specific objectives which you want to reach by using this summary. The summary is not just an enumeration of the work that you have performed; it should include the following:

1. The technical work that you performed. Don't go into detail on this. Assume that you're talking to people who have some knowledge of your area of expertise and just state what type of product or service you delivered.

2. State to whom you delivered your services.

3. State the date you finished the project and the time period over which the project ran.

4. State the dollar size of the contract. This is not always necessary, but it will often help your client to see the size of the contracts you are accustomed to handling.

5. State whether or not you met your schedule. If you came within the budget constraints of the contract, this should also be emphasized. Describe your services and how well they were received.

6. State whether or not you delivered what you promised. Include a letter from your previous client which states his satisfaction with your services. What you are trying to get across is the fact that you delivered what you promised. It's not your opinion that counts. What counts is the opinion of your previous customer. Also, you might want to emphasize that you're still dealing with this customer and that you still have a satisfactory relationship.

List of Publications

Some specialists have a long list of publications that they include with their resumé. This is incorrect; you should use a list of publications separately. A prospective client is only concerned with those aspects of your background that will help him solve his problems. You might be a Noble Laureate, but if you can't help him reach his goals, he will have little use for you. Therefore, make a separate list of publications and, after speaking with your prospective client and finding out what he wants, circle those publications on your list that show that you have experience in areas related to his problem. In some cases you might even want to send him a reprint. The point to be emphasized is that you should only stress those publications that relate to his project and objectives.

It's commonly believed that a specialist gets a reputation in his field by writing articles for trade magazines, technical journals, and newspapers. This is, of course, true. But there is one fact related to this that is often missed. The real value of those articles comes when you meet a prospective client who has never heard of you before and you show him a reprint of yours that is related to a problem that he is trying to solve. When this occurs, you can be assured that he will feel he has found the right person for the job.

Technical Reports

Over the course of your career, you will write a large number of reports. These reports can be used as future promotional material. If you write the reports in an incorrect manner, however, they will be difficult to use in this purpose. Therefore, when you're writing the reports, break each report into independent sections. Write the sections in a way such that the sections are as self-sustaining as possible. From each large report you can often make four or five small reports which could be used for promotion materials. After a while, you'll be sending these reports out like brochures. The people receiving them will appreciate the rapid and effective assistance. Even if the reports don't apply exactly to their problem, they will appreciate the fact that you cared about them and their project.

The technical reports which you send to prospective clients should not be long and detailed. Subsections of other reports that have the pertinent information are sufficient. Furthermore, if the person to whom you are sending the report wants more information, let him ask you for it. This opens the door for you to start interacting with him on a consulting basis.

This program can only be accomplished if you are organized. You should have a file drawer filled with technical reports which are available to be sent out, and you should have ready access to a complete list which describes the reports that you have available. Then, after speaking with the prospective client, you can tell your secretary to send him the appropriate reports. If well organized, this can all be handled without having to invest a large amount of time.

A Technical Information Pack

A technical information pack is a collection of promotion pieces put together in such a way that it applies directly to a prospective client's needs. When you're speaking to a person on the phone, write down the different pieces of promotional material suitable for his situation. You might be surprised that at the end of the conversation you have a list of six or seven items that your prospective client would find useful. Collect the necessary items and send them to him immediately. Include a brief letter with the information pack and follow up with a telephone call a couple of days later. You will usually find that the information was appreciated and that you have established a base of communication on which to form a future business relationship.

The important point which I want to emphasize here is that if you've done your homework and, over a period of time, created a large number of promotional pieces, the information pack which you send people can be well adapted to their specific needs. An information pack which has nonpertinent material is useless; in fact, it might even be harmful. But if you really understand what a person's problem is and then send him the proper information, you will be performing an excellent service and you will get a good response. Include the proper technical papers with the information pack. These can be papers written by yourself or people in your firm. Don't hesitate sending them copies of papers written by other people if they will help your prospect solve his problem.

Believe me, he'll return the favor sooner than you think. After all, you are in the business of helping people solve their problems, and as long as you are thinking about your prospects, you will be halfway home to having a successful business. This, of course, doesn't mean that you should spend all your time and money just helping other people without getting paid. That would be ridiculous. However, many of the activities which we are talking about take a small amount of time if you're well organized. Upon receiving this information, your prospect will feel that you've been thinking about his problem and he'll remember you in the future. He'll also be a source of information when you need some quick help.

Scale Models, Prototypes, and Flow Diagrams

If you're in the type of engineering or data processing business which builds anything that can be seen or touched, by all means have some kind of a scale model which you can bring with you to show to your potential client. If you have entered the negotiation phase with a potential client, you will find that a very effective means of showing him what you've done is to take some scale models with you and leave them with him overnight. If you bring four or five with you, leave one with each of the client's staff members and then, on the next day, switch the models around.

Let a model sit on your client's desk so he can look at it for a few moments every day. Leaving these models with your potential client will create an important impression. In particular, it gets the message across that you really do what you say you do. I can't emphasize how important this is. Leave these models with them, no matter how big they are, no matter how expensive. Some models such as models for buildings, bridges or apartment complexes are fascinating, and people like to have them in their office. These models can mean more than what you say in a sales presentation.

If a computer programmer has a notebook of simplified flow diagrams from some of his previous assignments, he can use them to show a prospective client how his problem is similar to those he has already solved.

BROCHURES

The objective of a brochure is very simple. It is used to communicate to large numbers of people and drive a high volume of inquiries into your company. The brochure should communicate your capabilities in such a way that the general reader can quickly get the message of what your company is all about and whether or not your services would be of benefit to a prospective client. If you are a one- or two-man operation, you probably will not need a brochure. A well-written resumé will suffice in this case. However, if your group consists of three or more specialists, a brochure can be used to describe your services.

Make It Brief and to the Point

Company brochures come in different sizes. Keep in mind that a brochure that is too large does not fulfill its objective. If you make a brochure too long, the chances are it won't be read, and it could possibly even confuse some of its readers. All you want out of a brochure is a quick statement of your services. This can easily be communicated in one or two pages or, at most, a five-page booklet. Any brochure that's longer than this is not needed. Once the reader contacts your company, he can then be sent an information packet with pieces of promotion that specifically relate to his problem. A brochure is not an independent promotion entity. It should be looked at as the first step in a series of promotion pieces. It's going to be sent to large numbers of individuals with the intention that they will make some kind of a response. Going into too great a detail is both unnecessary and extravagant.

Be Involved with the Design

In the initial development stage of the brochure, you should keep its design under your control. You and your associates are the people who know your public the best. You might have outside promotion people come in and try to tell you what your brochure should look like, but you and your associates should determine what information goes in the brochure and how it's designed. In designing any brochure you must keep the reader in mind, and you are most often the person who understands the characteristics of your prospective clients.

There are some people who believe that the design of a brochure is difficult because there is so much important information that has to be left out. How general should you make your brochure? You're going to lose sight of your main objective if you get into these problems. There's nothing wrong with a general brochure as long as it is followed with a company information pack, which is sent to a person who inquires about your company after he's read your brochure. A brochure does not need detailed information. State what type of work you do, and emphasize that you're experienced and that you can do the job. Just keep in mind that you're not running for mayor and that this brochure's objective is not to demonstrate your company's civic involvement or the like. The purpose of this brochure is simply and only to tell the public that:

1. You're out there and ready to do business;
2. You do a specific set of tasks; and
3. You're experienced and you can do your assignments damn well.

That's it. Don't get all hung up worrying about what kind of image you want to portray. The best kind of image comes from the fact that you promise to deliver a particular product or service and that you then deliver what you have promised. Tell them that you deliver what you promise. If you think that's being kind of bold, forget it. It's not being bold—it's being honest. Of course, if you don't deliver what you promise, it's a good idea to change your ways and start delivering better service.

When deciding on the kind of information to put in your brochure, you're going to have to face the fact that people just aren't going to read a long treatise about your firm. Stress the kind of information that lets the reader know that you can do the job. Stress your experience and avoid mentioning any references. Get the information across that yes, you're real, and you're really out there. It might sound silly, but that's an important message that you must get across to your public.

Small Outfits

If you're a small outfit that is just starting out, you're going to have to emphasize your specialty in the brochure. This is the aspect that shows that you can do certain specialized jobs that make you unique. Your uniqueness is a thing which has to stand out in the brochure, and this can be expressed by stating your specialty and who the key associates are who will get the job done. Though this holds for all companies, it is even more important for small companies, because in small companies it's just about all you

BROCHURES

have to offer. Let's face it, in a small outfit you haven't been around since 1895, and you're not going to be able to present an image of stability.

There are unique aspects to your firm that other firms just can't match. Find out how you differ from your competition and stress these points in the brochure. If you want to be seen as a consulting firm, you must include the unique aspects. These are the items which are going to catch someone's eye.

The Brochure on Brochures

I'd like to recommend a booklet, "The Brochure on Brochures," written by the American Consulting Engineers Council. This is a 25-page booklet that is an excellent introduction to every task you're going to have to do in getting your brochure ready. This is a booklet which is written for engineering companies and is written by engineers. It's excellent and I highly recommend it. You can obtain a copy by writing the American Consulting Engineers Council, 115 15th Street, N.W., Washington, D.C. 20005.

Write Your Competitors

Choose five or ten companies which do similar work and write them for their brochure. Take a look at all the positive and negative aspects of the brochures keeping their specialty in mind, and then write your brochure which emphasizes your unique aspects. People reading your brochure should be able to differentiate between the benefits obtained in working with your company as compared with other companies in the same field. Regard the other brochures as competitive promotion and then outdo them. It won't be difficult.

The introductory brochure should be given a lot of attention in terms of quality and appearance. This is the major item that will be reaching people and will be representing your company for the first time. Whatever is presented in this brochure will be regarded by others as the best job you could do in presenting yourself. After a company has been dealing with you and they know that you deliver what you promise, it won't matter if you occasionally send them a piece of material which is not layed out with extreme care. There are certain circumstances where this will happen, and your client will certainly understand. However, when it comes to dealing with somebody for the first time, formalities *do* count, and the formality of putting out a high-quality piece of promotion is important.

A Checklist for Improving Your Brochures

While you're preparing your brochures, you might ask yourself the following questions:

1. What is the major message being delivered? Is it the message that you want delivered? Are you stressing the relationship between the benefits, advantages and features of your service?

2. Is there credibility to what you're saying? You should seriously ask yourself whether the piece of promotion you've written will be believed by those who are reading it. It might have to be toned down somewhat if it is not going to be believed by those reading it.

3. Have you avoided long, awkward sentences and phrases? There's no reason to get verbose in any kind of advertising or promotion. You'll find that short precise phrases are sufficient and they'll get the message across.

4. Is the brochure a piece of promotion which your public is accustomed to reading? If it's considerably different to what they've seen, make sure you haven't gone too far overboard. If it's extremely different, they won't see it as being very real and might ignore it.

5. Is the quality of the print and paper stock you're using proper? Don't skimp to save pennies on the paperstock. A little quality here goes a long way.

6. Are the color tones pleasant, or are there strong contrasting features that make the overall appearance of the brochure unpleasant?

7. What do other people think of it as an advertisement or promotion piece? Pass it out to a large number of people and ask them for their response. Often their response is going to be considerably different than your opinion. And remember something—it's *their* response which counts, not your opinion. When it comes time for choosing what's right and wrong with regard to design, your own opinion can sometimes lead to trouble. Therefore, listen to what others have to say. If you show it to enough people, you'll get a good feel as to what's right or wrong with your piece.

8. Have you stressed the most important benefits in your headlines?

9. Have you supported your claim of your major benefits by stating advantages and features of your service underneath the benefit statement?

10. Have you stressed the consequences of not dealing with your organization? If a potential client stands to lose by not dealing with you, this should also be stated in your promotion.

11. Does it stress the fact that your firm will help a client meet his objectives and that you won't overwhelm him by stressing a "we know what's best for you" attitude.

3
Promoting Your Practice

The purpose of this chapter is to present you with many of the promotional techniques you can employ to obtain prospects. It is most important that you realize that promotion is not only advertising. There are many ways your services can be recognized by others. Such recognition will not take place if you take a passive attitude and wait for the phone to ring. When it comes to generating new business, don't ask whether or not you should promote yourself. Accept the necessity of promotion as a stable point of reference and inquire about the promotion methods that are best suited for your practice.

REFERRALS

The most common method of getting new business is through referrals. After you have been in practice for a few years, prospective clients will hear about you from your old clients and business associates. In a similar fashion, you can ask some of your professional associates if they would refer you to someone who could use your services. This is how many consultants get started and keep new business coming in.

The major advantage to promoting by referral is that when it can be used it is very effective. There is no better way to set up a solid relationship with a potential client than to have him be referred to you by someone whom he trusts. It should also be mentioned that there is a significant disadvantage to promotion by referrals, and that is that it is unpredictable. If you plan to run a private practice with the hope that the telephone will start ringing all by itself, I can guarantee you that you will eventually meet with disappointment. Predictable leads and new clients will come only from actively promoting yourself, not from passively waiting for new business to come your way.

When you approach a business associate and ask him if he knows of anyone to whom he could refer you, be sure you're not engaging in a light and social conversation. If you are going to have someone help you, your request to him must really register on him and cause him to take a close look at everyone he knows who could be a potential client. If you are sure he could give you some good leads and referrals, you might want to take him to dinner for the express purpose of getting new leads. If you ask someone very casually if they know anyone that might need your business, they most likely will casually yawn and say that they will let you know if they think of someone. Don't take a yawn for an answer. Tell them that you want to "talk business," and they will listen and help you out.

COOPERATIVE ADVERTISING

Some forms of advertising can be expensive, and there are methods to reduce the cost. For example, if you are a programmer in the business community, you might want to join forces with two or three other programmers whose skills complement your own. You can take out some advertisements in a trade journal or a local newspaper. A good advertisement will often generate more leads than you can handle by yourself, and if you split the advertising cost three or four ways, you will be able to get some good leads at a very low cost. It is not necessary to be in a partnership or a formal association with the other professionals with whom you are advertising.

If you have a large group of individuals, you might wish to employ an agent on a part-time basis to promote you as a group or individually. Contract engineering and job shops play this role and can be of some assistance. However, in the situations where you and your associates are highly specialized, you might find it more effective to employ a representative who understands your specialty or who has contacts with prospective clients. A person who represents you in a case like this will often receive 10 to 25 percent of any contracts generated. If the contract is extremely large, the percentage could be smaller. If he requires a retainer, you can split the cost of it between the members of your group.

TRAINING CLIENT'S PERSONNEL

As new discoveries, inventions, and technical developments occur, you will find that companies require someone to keep their technical professionals abreast of new innovations. For example, I know a digital engineer who occasionally gives courses in microprocessor design to employees of large aerospace firms in Southern California. Sometimes it is more efficient and less expensive to bring a consultant to a firm to train the employees than it is to send a large group of engineers to a university. Also, a college course is generally not as practical as a course designed by a consultant that is geared specifically for the customer's needs.

The principal advantage of giving such a course is that it will put you into contact with a large number of technical peers and prospective clients.

I therefore recommend teaching a course such as this for promotional reasons; I have seen it done successfully many times.

A course also gives the students a chance to ask questions that they would be hesitant to ask outside the classroom. Providing the opportunity to expose one's ignorance gracefully is a crucial issue in dealing with prospects, and it is a sensitive issue for many people. An in-plant course is a good opportunity to introduce new concepts to prospective clients without embarrassing them.

PROMOTING THROUGH FIELD REPRESENTATIVES

A professional field representative is a person who independently represents a number of technical companies that have their home base a considerable distance away from the territory covered by the field representative. The field representative can play many roles for the company he represents. He can be a salesman, a repairman, a handler of customer complaints, etc.

Since he is often not highly technically trained, but is usually the first to find out if something goes wrong with the equipment he represents, and since the field representative is in contact with a large number of companies in his territory, he is often a good candidate for new business leads. If you can assist the field representative with problems that are too difficult for him to solve, you can be sure that he will use you as often as possible. Try to find out who the representatives are in your own field of interest and contact them to see if they can give you a few leads. In return, you can offer to assist them with any questions they might have that are of a technical nature. Many representatives are salesmen with a weak technical background, and they will welcome having a technical associate.

NEWS AND PRODUCT RELEASES

News releases and new product releases are an excellent way to obtain free advertising. If you have written a report on a new technical product, a product brochure, a computer program, etc., you can write a news release for it and send it to a large number of technical trade magazines. You should include a photograph of the new product or brochure and ask them to run it in the New Products or New Literature section of their magazine. There are many magazines with high circulations to whom you can send the releases. Complete information on all the technical trade magazines can be obtained from:

>BUSINESS PUBLICATION RATES AND DATA
>Standard Rate and Data Service, Inc.
>5201 Old Orchard Road
>Skokie, Illinois 60077

If you are already advertising in one of these magazines, or if you are planning to advertise in the future, you can send the release to the magazine's sales representative and ask him to get the release placed for you. Most magazine sales representatives will be happy to help you out. I have placed over fifty news releases through these sales representatives. Because trade magazines are in the business of reporting new data and product information, they will usually be very helpful to you.

BOOK PUBLISHING

There is probably no easier way to break into consulting than by being an author of a book that covers a technical subject. There are many authors turned consultant who never even dreamed of becoming consultants when they originally published their book. However, after their book was published, they received so many requests for assistance from their readers that they began a consulting practice based on popular demand.

Many technical books only sell a few thousand copies a year and the royalties from such a small number of sales are consequently quite low. However, the consulting fees that can be earned through this marvelous method of promotion are considerable. If your background is such that you are able to write a book in your field of expertise, I would suggest that you give such a project extensive consideration.

Being an author can also assist you in other promotional activities. For example, if you have written a book on computer system design, then it will be easier to give seminars or courses to a client company. This is an example where one form of promotion makes it easier to create other methods of making your service known.

TECHNICAL JOURNAL ARTICLES

One of the better techniques for technical promotion is the writing and promoting of technical journal articles. Note that I said promoting the article. Just because an article is released in a journal does not mean that you are finished using it as a promotion piece. These technical articles are often not seen by the right people. You can actually use articles as promotion pieces and send them in the mail. This method is effective and it has many advantages.

The people who read your article see you as an expert, and they become oriented to your capabilities by applying your data to their problems. They are gaining an understanding of a topic based on your information, and they are already relying on your expertise. Furthermore, the technical paper you write might be a key to a solution of one of their problems. When this is so, they will be delighted to have the opportunity to meet you, and they'll go out of their way to get to know another expert in the field.

Despite these advantages, many technical professionals don't promote their articles after they are published. By promoting technical journal articles, you can obtain noteworthy results. Technical articles have formed the foundations for many experts to start their own consulting firms.

Another important fact with regard to the recognition that is gained from technical journal articles is that when it comes time to sell or contract your services, people feel more comfortable making a commitment to an expert. Even if they don't know you, they feel that if you're well published, they can rely on you and, in most cases, they are right. The understanding that they gain from reading your paper is not only technical. The fact that you have written the article shows that you have the expertise. When you are promoting your services you might find that one of the major difficulties to be overcome is the fact that people just aren't sure of your competence. When they can read some of the articles you've written, it will put them at ease, and this will make your promotion activities a lot easier. In fact, having a large number of technical articles coming from your firm can often be the most significant reason people will choose to do business with your company.

Consulting companies don't realize this, and they often feel it takes too much time to write a technical paper. Even a small company can get some of its technical personnel to write these papers in their spare time. The articles don't have to be sophisticated. All they have to do is impart information that is meaningful to somebody else. Remember, you're not trying to receive the Nobel Prize; you're just trying to give information to help someone else.

The technical level of your article is important. If the technical content is too high, then its use as a promotional piece can be somewhat diminished. The important features of this technical article can and should be summarized in such a way as to reach a broad public.

PROMOTING THROUGH TECHNICAL TRADE MAGAZINES

As I have said, the major limitation of technical journal articles is that the public reached is somewhat narrow. Generally, the people you're trying to address are not other engineers, but program managers and administrators. In contrast to journal articles, the trade magazine article can be an effective means of reaching program managers and administrators.

The purpose of the trade magazine article is different than that of a technical journal article. A trade magazine article will expose your general background (although it won't show your technical depth as well as a technical journal article.) It will give you the type of promotion which can hardly be matched by any other technique. Furthermore, the number of people it reaches is very large. Some technical magazines have subscription rates as high as 100,000. The proper sequence of steps you should take in writing a magazine article is as follows:

Establish a List of Possible Magazines

Make a list of all the magazines for which you could possibly write an article. Sources of these magazines can come from your associates and from the *Ayer Directory of Publications*, or *Standard Rate and Data*. Find out what magazines your potential clients are reading. One good way to do this is to contact your existing or potential clients and ask them what they read. You can be perfectly straightforward about it. Tell them that you're planning on doing a promotion campaign for your company, that you're going to write some articles, and that you want to know what they read. They'll be happy to tell you. You might be surprised to hear that they read magazines which are not even on your company's subscription list.

Contact Editors

With the list of possible magazines in hand, contact the editor of each magazine and ask him what his magazine public wants and what would make a good article. Explain that you're planning on doing a little bit of promotion and that you are anxious to write a good article. They'll be happy to give you some topics. They're aware that consultants write these articles for their own promotion. The editors often know what their public wants to read.

Avoid Duplication

Get the most recent issues of all the magazines in which you might publish and determine what has been written in these magazines recently. From these magazines you'll get an idea of what has been written, as you'll want to avoid any possible duplication.

Demographics

Write each magazine to get their demographics. From the statistics the magazine sends you, determine which magazine is most suited to your article and public. The reader circulation is not always the most important. What's most important is your particular public and what percentage of your public reads the magazine. Again, a survey of your clients would be beneficial in finding out what magazine is best. If you don't have very many clients, just choose a magazine that you feel will reach your prospective clients. The important thing is to publish an article—don't spend all your time worrying about which magazine is the proper one.

The Editor Again

Contact the editor and explain the type of article you'd like to write. Get his verbal okay. Also ask him to send you a writer's guideline for articles in his magazine. If he does not give you an okay, call the editor from the next magazine on your list. Follow up the telephone conversation with a letter describing the article in detail and get the editor's preliminary okay in writing. Also, ask him for a date that this article will be published.

Write the Article

Remember that magazine exposure can be very useful, especially when you're going to employ these magazine articles in your promotion later by mailing reprints out to prospective clients. Here are a few hints in writing the article that you might find beneficial for promotion:

1. Try to write an article which is not just one article, but a series of two or three articles that are spaced a month apart. Often the editors like this because it keeps their readers looking forward to the next issue. It also gives you a much larger exposure.

2. If possible, use multiple authors in your article. The advantages of this are many. For example, if one of the authors leaves your group, you still have one of the authors remaining as a creditable source of the article. Also, it exposes more of your associates which, in itself, is important promotion. It also helps in handling inquiries when people call in and wish to be connected with one of the authors of the article. You'll also find that writing articles such as this is excellent for the professional development of your associates; they will get some of the recognition that they deserve. This is very important to their careers, and they'll appreciate your helping them out in this respect.

3. When writing the article, write with a particular public in mind. Don't try to make the article so general that it applies to everyone.

4. While writing, don't get complicated or very technical. Define all your terms, and make sure the article is very clear and easily understood. You're not writing to technical people; you're writing to administrators and managers who are your potential clients. Make your messages short and to the point.

5. Use many pictures and diagrams. This makes the article much more interesting and gets the message across without unnecessary verbosity.

6. Mention your company's expertise as often as possible. Remember that this article is a type of promotion and that you should use it as such.

7. Include a photograph of yourself and a short biographical description which will be included with the article. Be sure to stress that you are a consultant.

Edit and Proofread

After writing the article, give a copy of it to a professional editor and have him go through it thoroughly. It will only take him a couple of hours, and it won't cost much. It could save you a month or two of wasted time due to publication corrections. After it has been edited, proofread, and corrected, send two or three copies to the publisher and get his comments. If everything has been done correctly, he should send you a notice of publication.

Reprints

With the article now in for publication, you should inform the editor that you would like to have as many reprints as possible. Have him send you the maximum number. Some magazines won't be able to give you a very large number of reprints. However, by paying an extra fee, you might be able to get a larger number of reprints very inexpensively. If not, you can have the reprints made up by a printing subcontractor. Don't photocopy these reprints to use them as promotion material. The photocopy will not come out well, and it's a better idea to go to a professional printer who can have the article printed.

Use the Magazine Cover for Promotion

In addition to the article, you should attach the cover of the magazine to the front of the article. This is extremely important. The magazine will be happy to supply you with all the magazine covers that you request. This gives them added visibility and, of course, it adds some reality to your article. If you can't get a large number of magazine covers from the magazine publisher, then go to an independent printer and have him print the covers. It will not be difficult to have this done. The magazine will give you price quotes on extra magazine covers, and you can compare this to what it would cost to have a printer of your own choice do the job.

Send the Reprints to Everyone

When these reprints are available, send them to absolutely everyone. You should have no doubts or reservations about sending these reprints out. Send them to everyone. Include them in letters. Include them in inquiries that came into your company over the phone. Include them in information packs on your company. Include them with the company brochure.

PLACING A MAGAZINE ADVERTISEMENT

Here is the correct sequence of steps to take in creating and placing a magazine advertisement:

Identify Your Public

Who are the people who will be contacting you first as prospective clients? Often the program manager will be the one who signs on the dotted line, but it may be a technical professional who makes the initial contact. In this case it's the technical professional who makes the initial contact that would be considered your prospect. If you've had a few years of experience in engineering or data processing, you should have a pretty good idea as to who this is. Write all of your different types of prospects down on paper. Don't walk around with a fuzzy idea of who your prospects are.

Find Out What Magazines and Publications They Read and Review Them

The simplest way to do this is to get on the phone and survey them. Tell them exactly what you're planning. You might be surprised when they mention magazines that you have never read. Make a list of the magazines mentioned most often.

Review the magazines. If you're already subscribing to some of the magazines, then you're halfway home. Determine which magazines are pertinent to your work and subscribe to them. Many of these magazines don't cost anything to subscribe to. All you have to do is be in the field, and the trade magazines will put you on their mailing list and let their advertisers foot the bill.

Write the Advertising Editor or Sales Representative

Do this for each magazine and get complete statistics on the subscribers. Determine the percentage that can be considered your public. You're also going to need information on the cost for placing different sized ads. Ask them to send you the magazine's media pack which will contain all the rates.

Be sure to ask what the restrictions are with regard to classified ads and what the price per word is for such ads. Classified ads are important and often inexpensive.

Prepare Your Ad

Before putting the ad together, call ten potential clients or people who hire consultants and ask them what benefits they want to receive by hiring a consultant. Write these down. If you can, try using these in your advertising as the benefits that you offer.

Send a Typed Copy of this Ad to an Ad Agency

Contact a local ad agency and tell them you would like them to place an ad for you. Give the name of the magazine, the details, and the size of the ad which you want, and tell them to take it from there. They usually charge a standard fee that's a certain percentage of the cost of the ad. It's usually not very much, and it's well worth the price.

Give them a 50% advance for the ad and tell them that you'll pay the rest when you see the ad placed correctly and on time. If you haven't dealt with them before, they might want all the money up front in the beginning. Although this is not advantageous, you might not have any choice. Once you've established a good working relationship with them, give them the 50% advance and make sure they place the ad correctly. Remember, any mistake which is made or any time delay will cost you money. If it's their mistake, they are going to have to pay because they didn't deliver what they promised. It's as simple as that. Of course, if you are only going to place classified ads, you won't need an ad agency.

If You Are a Specialist

Instead of advertising in the technical trade magazines, you might consider putting some advertisements in the research and development journals of your trade. Keep in mind, however, that although you might read these journals, the people who hire you might not. You're going to have to determine if the readers of these technical journals are really your public.

Stress Benefits

When writing an advertisement, you will want to stress the benefits in dealing with your firm. I have made a list of the objectives that are common to many prospective clients. You can use this list in writing your ads and when speaking with a prospective client.

1. To have certainty in the outcome of his project.
2. To get repeat business.
3. To save money.
4. To get more business.
5. To have a more effective staff.
6. To meet project deadlines.
7. To have more time in the work day.
8. To have equipment that he can rely on.
9. To have manpower and suppliers on whom he can rely.
10. To get respect from peers.
11. To advance in position.
12. To have the project succeed.
13. To understand and obtain the solutions to present problems.
14. To have expertise in an area.
15. To have someone who can be relied upon and who will be available when needed.
16. To have a subcontractor who will deliver what was promised on time.
17. To have more order and predictability.
18. To have more energy.
19. To have more cooperation.
20. To have improved communication.
21. To get more enjoyment out of his job.

USING THE CLASSIFIED SECTION OF TRADE MAGAZINES

An excellent source of advertising which is often neglected by small companies and consultants is the classified section of trade magazines and journals. One of the reasons why the use of the classified sections is so effective is that employers often look in this section for people who can help them out on a particular problem. Sometimes they are looking for a full-time employee who specializes in a particular area and they are having a difficult time finding such a person. When they see your advertisement in the classified section, there is a probability that they will get in touch with you. Don't worry about the probabilities—the cost of these classified sections is so low that even if you received only one return a year it would probably be worth it. The subscriptions on many of these magazines is 10,000 to 100,000, and the exposure that you receive for the money is excellent.

Your classified ad doesn't have to be long or very fancy. Just state your specialty, your name, address and telephone number. Place the ad in the trade magazines that are read by the people who are most likely to need your assistance. Remember—these people are often managers, not highly-skilled engineers and data processors. Therefore, they may not be subscribing to the same trade magazines as yourself. Pick the magazines that the managers and technical professionals read.

ONE-TO-ONE TECHNICAL PROMOTION

For any firm or individual who works in a technical specialty, there exists a very specialized form of promotion that obtains excellent results. It is ideal for someone such as a consulting engineer or a data processing professional. In fact, using the following technique, you can generate more contracts than you really know what to do with. However, it does require that you be a specialist and that you interact with specialists. The steps are as follows:

1. Make a list of all the research journals, technical journals, and trade magazines which apply directly or indirectly to your area of specialization.
2. Get copies of each publication dating back at least two years. If you don't have immediate access to these journals, then you can get them at the nearest technical library.
3. Go through each journal and mark each article that applies to your specialty.
4. Write a general letter to the authors. An example would read as follows:

Dear _____

   I found your article entitled, "_____" interesting and I would appreciate it if you would send me a reprint and copies of any other papers or reports that you've written in this area.
   I'm enclosing a brochure which describes our activities and, as you can see, there's some overlap of interest; so I'd like to stay in close communication with you. Please keep my name on your mailing list and keep me up to date on your activities.

                                                  Best regards,

5. Send this letter to the authors of the journal articles. If there are hundreds of people to send it to, send it to hundreds. Don't fret. That's a very small price to pay. Even if you have to put this letter on a computerized typewriter, do it, and get it out. It's going to cost you a very small amount of money for each letter, and you will get responses from most of the authors.

6. When you receive the reply, go through it with the following two points in mind:
    a. Review it to see what technical points you have in common with the respondent.
    b. Review it to determine if in the future you might be able to use the author as a potential professional associate. Determine if there is any possibility of your interacting with him on a business basis.

7. If a respondent looks like a hot prospect you should contact him immediately. Write him a short note including any reprints or reports in which he might be interested. If you live in a metropolitan area which holds conferences, invite him to your place for lunch when he's in the area, or make plans to meet him at the next conference. With this technique you can get together with five or ten new prospective clients at a three-day technical conference.

8. Whenever the opportunity arises, write him and ask him his opinion as to a particular problem on which you're working. You're doing many things at once here. You're establishing a better communication with him, as well as giving him an opportunity to voice his opinion on how something should be handled. People like to be asked their opinion in various matters. This will be an opportunity for him to help you out, and you'll be able to do the same for him with regard to his problems. Before you know it, you'll have a strong working relationship. After you have established a strong relationship, you won't have any trouble extending it into a business contract.

HOW TO PROMOTE YOUR EXPERTISE WHILE ATTENDING A SEMINAR

1. A good way to promote during a seminar is to contact the person who's giving the seminar in advance to get a list of people who will be attending. Note the people on the list to whom you'd like to promote, and contact each before you leave for the seminar. Mention to them that the seminar speaker mentioned that they were coming and that you were looking forward to meeting them at the seminar. Speak to as many people as possible before you go to the seminar. Don't get into details prematurely—just call them on the phone, introduce yourself, and ask them about their major interests and technical problems. Recognize that you have much in common with these seminar attendees; if they are attending the same seminar as you, there will be an overlap in interest. They can also be sources of names of people within their own company.

2. When you get to the seminar, get a list of everyone attending and, next to each person's name, write down their interests. Pick the people who have interests close to yours and speak to them during the seminar breaks. Introduce yourself and tell them that you'd like to meet with them at the end of the seminar for lunch. Set up private appointments with as many people as possible.

3. Avoid going to lunch with crowds of seminar attendees. If you do, recognize that it's going to be rather poor for business. You might just want to have lunch "with the guys." That's fine, but realize that having lunch with the guys is a very ineffective means of communicating with someone. The best way to promote during seminars is to one person at a time.

4. Participate in the seminar. Listen, ask questions, and pay attention to questions the prospective clients might have. Do what you can to make the seminar a success in every way you know how. If someone asks a question that is related to your interests, go up to the person afterwards and talk to him about it. A seminar is not only a time for data collection—a seminar is an excellent opportunity for you to get to know new people in your field.

5. When you are speaking with someone, be sure to find out where they need assistance. If they need help in a subject where you are knowledgeable, be sure to tell them you can help them meet their objectives.

HOW TO PROMOTE AT A TECHNICAL CONFERENCE

Your objective at a technical conference is to meet privately with as many prospects as possible. You should go through the following steps in preparation for the conference.

1. Make conference reservations well in advance. Make reservations in a hotel that is as close as possible to where the conference is being held, preferably in the same hotel. When making your reservations, make sure you get an excellent room, and demand from the hotel that you know your room number in advance. You're going to be using your room number in setting up appointments before you get to the conference.

2. Get a conference schedule as early as possible and make up your daily schedules by determining what lectures you wish to attend.

3. Write or call everyone on your potential-client list to see who's going to the conference: Your list should be extremely long. Tell your prospect that you'd like to meet with him at the conference and set up a time to get together.

4. Set up an appointment sheet that is broken down by the hour. Make sure you see everybody in your hotel room. Don't go wasting your time or energy running around looking for somebody.

5. When people start calling you back to confirm appointments, set up each appointment by the hour. The most important prospects should be set up for the breakfast, lunch or dinner hours. If it's a three-day conference,

you've got opportunities for appointments for 9 or 10 meals and plenty of meetings in between.

6. Bring large stacks of all your promotional materials—your brochures, reports, technical papers and product samples. Set them up neatly in the room, and before you meet with each person, decide what pieces of promotion you want to pass along to him. Make a list of your objectives and know exactly what you want to communicate in your meeting.

7. In the morning, very early, call his room and confirm his appointment with you for that day.

8. See him for 45 minutes or less, and use the remaining 15 minutes in preparation for your next potential client. *When you meet with him, find out what his problems are and tell him where you can assist him.*

9. When attending a lecture on topics that are related to your interests, ask the chairman of the lecture (or the speaker) to get a list of everyone in the room so you can contact them later.

10. If you are giving a lecture, pass out a sheet of paper for the attendees to write their names and telephone numbers on so you can contact them afterwards. This is very important because the people who are attending your lecture are prime candidates for future work.

HOLD YOUR OWN SEMINAR

While attending your seminar, the potential client has an opportunity to get to know you over a period of time. A positive point about a seminar is that it allows a prospective client to recognize who you are and to take a long look at you and your capabilities. It gives him an opportunity to get into communication with you.

As the seminar progresses, he'll be able to see how his problems relate to the various topics in which you are knowledgeable. Sometimes potential clients consider that they are not ready to talk to a specialist, and that they wouldn't even know what questions to ask. Sometimes he will feel uneasy about his ignorance. The seminar provides an arena for both of you to gain an understanding of each other *slowly*; an understanding is gained over a period of many hours and even days. Finally, at some point in the seminar, the prospective client is going to realize, "I can deal with this specialist. After all, he's human, and I have no fears of expressing my ignorance. There are even a lot of things he doesn't know that I do." At that point he realizes that he can deal with you, and you are in a position to establish a good relationship with him. Remember that not all prospective clients are bursting with self confidence.

The point at which people reach this stage differs depending on their own technical background and their own personalities. The important point to realize is that this step *is* reached by most people at the seminar, and that it happens at different points for each individual. Sooner or later each person gets to the point where he feels competent enough to interact with a specialist. After that point, he will feel free to make a commitment to invest the necessary time and money to have you assist him in solving his problems. Once you have a person who is comfortable with you, you can deal with him on a one-to-one basis and shift into a highly productive working relationship.

4

A Promotion Program for Your Practice

Most of you have already had years of experience in your own fields and have become quite successful by knowing how to choose the proper people with which to work. The same is true in operating successfully as a consultant—you must know how to choose your clients. Note that I'm stressing the fact that *you* are to choose your clients. It's not the other way around. Your route to success as a consultant is to create a program that develops so many prospects and potential clients that you will be able to pick the cream of the crop and only work for the very best of clients. Not only is this the most prosperous route, but it is also the most enjoyable.

The objective of this chapter is to give you a specific plan to obtain a large number of clients. There are dozens of books and reports about sales—books with gimmicks, tricks and every game ever invented. This chapter is a sales plan--a straightforward set of steps. It doesn't have a trick in the lot. In fact, once you see your sales on the increase, you'll realize that this plan is just common sense.

YOU DON'T SELL TECHNICAL PRODUCTS OR SERVICES LIKE YOU SELL CAMELS

Ahab:	"But this camel requires less water than yours. Think of all you'll save."
Moham:	"True, but water's cheap."
Ahab:	"But compare his coat to the coat of your camel . . . ."
Moham:	"Forget it Ahab, I don't need another camel. What I need is an audience with the Princess."
Ahab:	"No doubt with this camel you'll catch her eye."
Moham:	"Hmm. What are you asking for it?"

When a guy buys a camel, it could be for any reason under the sun. And as we can see from the anecdote above, that reason is often created by the salesman. Not so in engineering and data processing!

Technical clients know what they want, but by and large, they don't know what's available on the market. Why? Because the marketplace hasn't gotten in touch with them. And that's because the technical marketplace doesn't know *how* to get in touch with them. In fact, when a technical professional sees that your service will assist him in reaching his objectives, he doesn't have to be sold on anything. *He* takes over and buys.

WHY IT'S EASIER TO SELL TO A TECHNICAL PROSPECT

The technical sales prospect is very different from the general buyer in many respects, all of which make it easier to sell your service. Here are a few characteristics that differentiate the technical client from the general client. Though these characteristics are not present in all technical clients, it is certainly true that more are present in the technical professional than in any other type of client.

1. *He's aware that he needs assistance* and knows that somewhere in the marketplace there is someone who can help.
2. *He has specific objectives* that some technical service will help him reach. He often has had experience with similar service.
3. *He's willing to listen* to anyone who will help him meet those objectives.
4. *He has the ability to see* that some services just won't do the job and will appreciate any service that can help.
5. *He has the ability to judge* how your service will fit into his overall plans. He has a good feel for where he's been and where he's going with respect to his project.
6. *He understands his specialty* and will understand the benefits he will receive from your services if those benefits are presented to him correctly.
7. *He doesn't have to be convinced.* Once he finds the right product or service, he will want it.
8. *He often has the funding to buy the service.* For the right service, money is not a major issue. If he doesn't have it, he can often find the funding.
9. *He'll make adjustments* in his budget and schedule to accommodate the consultant. He's also willing to negotiate.
10. *He is not apt to balk* at the last minute when there are agreements or contracts to be signed.
11. *He is realistic about delivery dates* and doesn't expect the world from you overnight.

12. *He is aware that things can go wrong* with the service and is willing to correct it with you.
13. *He's willing to share credit with those who helped him succeed.*
14. *He usually pays his bill!*

From this list you'd think I'd have a technical customer doing everything but giving the Sermon on the Mount. But I'm serious! I can think of no field other than engineering or data processing where one finds such knowledgeable, aware and cooperative customers.

"All right," you say, "then why in God's name is it difficult to sell our technical service?" *The answer is that you don't have large numbers of qualified prospects.* If you did, the sales would almost take care of themselves. The answer does not lie in sales. The answer lies in a successful marketing campaign which drives in hundreds of prospects from which you must *choose* the most *qualified* prospects.

PROFESSIONAL PROMOTION STEP BY STEP

Amateur technical promotion is very common. A company places a few ads, hires a salesman and calls it promotion. Well, it's not. It's not even close.

This chapter is going to show you how to promote yourself and drive in more requests for your services than you could possible use. The plan outlined here is not a vague generality on how it might be done. It's a step-by-step procedure which shows you exactly what you have to do to reach your sales objectives.

Step One—Sales Objectives and Planning

There are two things that you will accomplish by doing this step. First, you will determine your objective with regard to your future sales. This is the income that you want to have by a particular date. The second part of this step is to have you set a quota of sales presentations, or job interviews, that you will have to make in order to achieve the income that you determined. As you could have guessed, there is a direct relationship between the number of sales presentations you make and your total income.

Gross Sales

Your first step is to write down exactly how much you want to earn in gross sales over the first 12 months. Write it down here:

Gross Sales =

Contract Size

The next step is to write down your average sale or the average size of a company contract. Some technical companies have a very broad range in the size of their contracts. It really doesn't matter. Over the past years you've had a certain number of contracts of an average size. If you're just starting out, pick a dollar value of an approximate contract.

$$\text{Average Contract Size} =$$

Number of Contracts

The third step is to determine how many contracts you need. That's simply your first goal divided by the average size of the contract.

$$\text{Number of Contracts} = \frac{\text{Gross Sales}}{\text{Average Contract}} =$$

Presentations

Determine how many presentations or job interviews you will have to make to get this many contracts. Make a review of all the sales presentations you've made and find out how many presentations you have to make before you make a sale. Call that number the Probability of Contracting, P_c. The number of sales presentations that you must make is equal to the number of contracts that you must acquire divided by the Probability of Contracting.

$$\text{Number of presentations or interviews} = \frac{\text{number of contracts}}{P_c} =$$

If you're new in the business and you haven't had very many interviews, then you're going to have to pick a figure that you consider reasonable for your probability of contracting. Don't worry about guessing correctly now, because all we're doing is getting a gross estimate for the whole program.

Inquiries from Qualified Prospects

Note that before you give a sales presentation to anyone, you should be certain that the person is qualified. Do not make the mistake of wasting time and money on a sales presentation before you determine whether or not the prospect is qualified. Not all inquiries coming in are inquiries from qualified prospects. The question here is how many inquiries do you need to come into your company in order to obtain this number of qualified prospects. At this point we've arrived at an unknown, but for estimation purposes, I want you to make a guess at this. This is going to vary quite a bit from consultant to consultant. In some companies it might be one out of four. In some companies it might be one out of two. Nevertheless, you've got to put down an estimate of what you think it's going to be.

No. of Inquiries (to obtain one qualified prospect) =

Now we have the equation:

$$\begin{bmatrix} \text{Total No. Inquiries} \\ \text{(to generate} \\ \text{determined} \\ \text{gross income)} \end{bmatrix} = \begin{bmatrix} \text{No. of Inquiries} \\ \text{(to obtain one} \\ \text{qualified prospect)} \end{bmatrix} \times \begin{bmatrix} \text{No. of Presentations} \\ \text{Necessary (assuming} \\ \text{every qualified} \\ \text{prospect receives} \\ \text{a presentation)} \end{bmatrix}$$

PROFESSIONAL PROMOTION STEP BY STEP

Cost

The next question is how much promotion do you have to do to get this number of inquiries? As an example, let's take a small consulting firm that demands $200,000 in gross sales for the next twelve months. A review of their contracts shows that the average size of a consulting contract was $40,000. This requires that they obtain at least five contracts over the next year. If the probability of contracting for this company was 0.5, they would have to make 10 presentations/interviews to reach their goal. Therefore, they would have to have 10 qualified prospects before they could even hope to gross $200,000.

If this company had three inquiries for every qualified prospect, it would then require 30 serious inquiries every year before it could reach a sales goal of $200,000 per year. If further investigation shows that this company required 50 pieces of promotion to be mailed out for a serious inquiry to be generated, then for this company to reach $200,000 in sales per year, it has to mail out a minimum of 1500 pieces of promotion a year to obtain its sales goal.

If the average price of each piece of sales promotion is four dollars (which includes labor and all production costs) then the total cost for the mailing would be $6,000.

Savings

Let's see if the $6,000 invested in advertising was worth it. Assume that your potential gross income is $200,000 per year. Further, let's assume that you do not work 10% of the time because of periods of no business due to lack of contracts. If your $6,000 advertising program reduces your slack time from 10% to 5%, this will mean a $10,000 savings, which will more than pay for your advertising costs. Moreover, your company will have set the foundation for further repeat business.

Make Your Own Estimate

First make a gross estimate of what you're going to need for promotion. Go through your own set of numbers as above, and get a rough idea of how much you'll have to spend on promotional material and mailing.

Step Two—Creating Mailing Lists of Possible Clients

As you saw in Step One, the essential element of successful promotion is getting your message out to so many people that the sheer enormity of the promotion alone will guarantee you a more than ample supply of inquiries.

The objective of Step Two is to obtain a list of hundreds of individuals to whom you can send your promotion with the objective of obtaining qualified prospects. Now, when I say hundreds of people, I'm not kidding. In the sections that follow, you will be shown how to prepare lists which have large numbers of people who are directly related to your technical service. One of the unique aspects of the technical profession is that it's possible to generate lists of thousands of people who are directly or indirectly interested in the service which you have to offer. I know of no other sales area where it is so easy to make such a long and detailed list of prospects.

Remember that your objective is to bring in a large number of inquiries into your company. Only a small percentage of these inquiries will lead to qualified prospects. Thus, you're going to need a large number, and *your program is going to be as successful as your mailing list is large*. A large percentage of your new business can be generated in this fashion.

Commercial Mailing Lists

One thing which you will have to learn is the technology of mailing lists. Since the advent of the computer, the formation of mailing lists and list companies has grown rapidly. In addition, a mailing technology has been developed such that you can locate the particular public of your choice with extreme accuracy.

For example, if you're involved in the computer industry, you won't want a list of everyone involved with computers. That's not your public—it's too general. If you specialize in microprocessors and microprocessor specialists are your public, you can find computer lists which have a subsection that will deal with people who only specialize in microprocessors. Furthermore, many computer lists are broken down by zip code. Therefore, if you are interested in a particular public in your own area, you can ask the list company to give you a listing of these potential prospects by their location.

Technical consulting companies rarely use mailing lists to their advantage, and many consulting companies are not even aware that they exist. Most companies don't even have a well-thought-out marketing program, let alone one which is based on driving a large number of inquiries into their company. This is an advantage which you have over others. Use it. Without a large list of people to whom to promote, it's very unlikely that you'll be able to produce a large number of inquiries. You want to pick and choose those with whom you'll be dealing. Don't put yourself in the position of hoping for clients. Let them hope to get you as a consultant.

Lists of engineers can be obtained in a publication called Direct Mail Lists—Rates and Data, published by Standard Rate and Data Service, Inc., 5201 Old Orchard Road, Skokie, Illinois 60077. They publish a catalogue which contains thousands of lists, each list containing thousands of people.

Sources for Engineering and Data Processing Mailing Lists. The following is a list of a few companies which have large lists that can be rented. This is only a sample of the companies available. You can obtain much more information by subscribing to the Standard Rate and Data directories, or by getting them at the library.

ALVIN B. ZELLAR, INC.
475 Park Avenue South
New York, N. Y. 10016
(212) 689-4900
(312) 372-6880

CAHNER'S PUBLISHING COMPANY
5 South Wabash Avenue
Chicago, Illinois 60603
(312) 372-6880

BARKS PUBLICATIONS, INC.
400 N. Michigan Avenue
Chicago, Illinois 60011
(312) 321-9440

BENWILL PUBLISHING CORP.
1050 Commonwealth Avenue
Boston, Mass. 02115
(617) 232-5470

CLEWORTH PUBLISHING CO., INC.
1 River Road
Coscob, Connecticut 06807
(203) 661-5000

CMP PUBLICATIONS
280 Community Drive
Great Neck, N. Y. 11021
(516) 829-5880

DIRECT MEDIA INCORPORATED
90 South Ridge Street
Port Chester, N. Y. 10573
(914) 937-5600

DUN & BRADSTREET, INC.
Marketing Services Division
National Sales Dept.
99 Church Street
New York, N. Y. 10007
(212) 285-7468

INSTITUTE OF ELECTRICAL AND
ELECTRONICS ENGINEERS
345 East 47th Street
New York, N. Y. 10017
(212) 644-7768

McGRAW-HILL, INC.
1221 Avenue of the Americas
New York, N. Y. 10020
(212) 997-2377

COLLEGE MARKETING
GROUP, INC.
198 Ash Street
Reading, Mass. 01867
(617) 944-7515

COMPILERS PLUS, INC.
295 Madison Avenue
New York, N. Y. 10017
(212) 736-2288

DEPENDABLE LIST
MANAGEMENT
257 Park Avenue South
New York, N. Y. 10010
(212) 677-6760

RESEARCH PROJECTS
CORPORATION
Executive Plaza
50 Clinton Street
Hempstead, N. Y. 11550
(516) 481-4410;
(212) 895-1048

SUTTON PUBLISHING
COMPANY, INC.
172 South Broadway
White Plains, N. Y. 10605
(914) 949-8550

TECHNICAL
PUBLISHING COMPANY
1301 South Grove Street
Barrington, Illinois 60010
(312) 381-1480

UNITED TECHNICAL
PUBLICATIONS
645 Stewart Avenue
Garden City, N. Y. 11530
(516) 222-2500

ZELLER & LEITICA, INC.
15 East 26th Street
New York, N. Y. 10010
(212) 685-7512
Toll Free: (800) 221-4112

NATIONAL BUSINESS LISTS, INC.
162 N. Franklin
Chicago, Illinois 60606
(312) 236-0350

What to Ask List Companies

If you decide to rent lists, you should ask list companies to provide you with the following information:

1. What company controls the list? Find out if the company from whom you are renting the list is a broker for another company who actually controls the list.
2. Do they have branch offices in your area?
3. A complete description of the list telling you who the listees are and how they are separated in categories of their specialty.
4. The total list size and its rental rates.
5. Any information with regard to commissions and credit policy.
6. Information on their delivery schedules. The lists come from computers; it will take them a few days to retrieve the information.
7. Information on any restrictions that accompanies the list. For example, some lists may not be used more than a few times.
8. Can you test the list? Many of the list companies will make arrangements with you to sample a small portion of the list. That is, you can mail to a small number of names to determine if a mailing to a larger number would be profitable.
9. Some list companies also have letter shop services. If you give them the material you want mailed, they'll take over from there and fold, stuff, seal the envelopes and put them into the proper mail bags and mail them to your prospective clients.

Creating Your Own List of Potential Clients

There are some disadvantages to rental lists. One is that there's a delay from the time you rent it to the time that you actually get your promotion material mailed out. A second is that the list that is rented will contain many inappropriate names. Therefore, you're going to want to create your own list in addition to renting a list. The way to start is to create a filing system based on three-by-five inch cards and begin collecting names and addresses from everywhere. Here is a rundown of the possible ways of creating your own list.

Write down everyone you know in your industry. This is going to have to be an ongoing project, because you'll be thinking of different names as you go along. You might have a list already if you've been employed for a while, based on telephone numbers which you have collected in dealing with people. However, now you're going to have to think of as many people as possible. One efficient way of doing this is to simply sit down with a dictation machine and start thinking of various names. Have your secretary or an assistant get their addresses and phone numbers placed on 3x5 index cards.

Names from Partners and Associates. Get names from your working associates. Have each of your associates create his own list and supply it to you. Each employee or associate can supply a rather long list of names—a minimum of 20 or 30. This can expand the company list considerably. You

PROFESSIONAL PROMOTION STEP BY STEP

can obtain many names by finding a technical professional in your area of expertise who is willing to generate a list of his own and having him give it to you. If you explain to him in the beginning that you're not going to be using his name as a reference, he'll be willing to give you his names and addresses, provided he knows you'll be using them with discretion.

Company Personnel Directories. The employee directories of the companies in your industry will be an excellent source. If you're not sure which employees in the company are directly related to your kind of work, you can go through the directory and look for the names which you recognize and check them off. Then, after you've finished, hand the directory to your associates and have them do the same thing. You'll get quite a few names this way, and you will find that it's quite easy.

Industry Directories. One of the most accurate sources of lists are published directories of companies in your area of specialty. These directories are one of the mainstays of communication for people in various specialties. An industry directory should have lists of company presidents, vice presidents and marketing managers. Some directories even give the names of technical managers. These people should definitely be put on your list. If you do a thorough investigation, you might find that there are many industry directories in your area of specialty, and you will have to go through each one of them. Hundreds, or even thousands, of names can be obtained this way.

Yellow Pages. Go to the library and get the phone directories for the major metropolitan areas in the United States and look up your area of specialization. For example, if your clients are structural engineering firms, look up "Engineering" in the Yellow Pages and then locate the "Structural Engineering" companies.

Write Companies. If you have the names and addresses of the employees of some key companies, you should write each of them a letter. Get the names of the president, vice president, directors of marketing, data processing and engineering. Even if you don't know any employees, you can still write them. Some companies will be happy to give you this information, for they will consider you a new source of business.

Advertising Returns. After your advertising campaign gets going and you start getting a large number of inquiries coming in, you'll be able to use the returns from your advertising to add to your list. The best kind of promotion for a list formation is advertising in technical magazines, whereby a reader can circle a reader response card. The magazine then sends you a listing of all the people who are interested in your advertisement. Most magazines will send you the names and addresses of the interested subscribers on labels, and you can just transfer these labels onto your 3x5 cards.

Magazine Article Responses. If you write for magazines or technical journals, you can expect replies and inquiries to be generated by these articles. These names are important and definitely should be included in your mailing list.

List Companies. You should inquire to as many mailing list houses as possible and find out what types of lists they have and see if you can buy an entire list. Some companies will sell lists instead of renting them, and a thorough investigation should be made to find out if your potential clients exist on a list that could be bought. If this is the case, it could save you considerable time and money.

Authors of Magazine and Journal Articles. Collect all the trade magazines and technical journals in your area of specialty and go through each one of them. Write down the names and addresses of all the authors who specialize in your field. Often these authors are leaders in their fields, and many of them will be interested in your promotional material.

Technical Societies. One of the reasons for a technical society's existence is to get you in communication with their members. Write the technical societies to which you belong and see if you can obtain a membership listing from them. There are many technical societies which publish an annual list of all their subscribers, and some of these society lists also tell you what areas of interest the various society members have. The technical society meetings are another excellent source of names of people who are actively interested in various aspects of your area of specialty. Write to the society and inquire if they will sell, rent, or lend you a list of people who attended their recent meetings.

Government Contract Sources. The government prints publications of their contract awards. Also listed is the awarding government agency and an individual in that agency who is responsible for the contract. Determine what government publications would be useful to your business and subscribe to them. Then, each month, go through them and collect the names which would be useful to your business and add them to your list as potential clients.

Government Directories. There are also government directories of the laboratories and various agencies. Make a list of all the government agencies with which you could potentially do business, and write to them. Get a listing of their administrative structure and find out who should be receiving your promotion.

Department Lists. If you deal with clients who are in large companies with large departments, you should obtain department lists from the program managers. You can also ask them for an administrative organizational chart to show you who has particular responsibilities. From the organizational chart and the department employee lists, you should be able to determine who should be on your mailing list. A company phone directory is also very helpful and some companies will send you a copy upon request.

Marketing Departments. If you write to various companies which might be potential clients to get their product and service information, you will be able to obtain the names of people in their marketing departments. The salesmen will help you locate potential clients in the company. It's easy to get information from people in the marketing department if you don't know anything else about the company. For example, if you wish to find out who the program managers are on various projects, just call one of the marketing people and he'll tell you. Salesmen are often anxious to open any kind of communication with a potential client, and they'll provide you with as much information as possible. You don't have to pretend that you're going to buy one of their products just to get some names. If you're very straightforward about it by telling them that you're interested in doing business with their company and that you don't know who to talk to, you'll still receive a favorable response.

Technical Publications. Write to the distribution department of technical magazines and ask them if you can rent their subscription list. It could be a source of thousands of potential prospects.

Add to Your List Daily. You should make it a habit to write down the names of everyone with whom you become involved. After a meeting, make a list of everyone in attendance and give this list to your secretary. She can get in contact with the people by phone and get their addresses and any other pertinent information.

Hiring A Competent Mailing House

Mailing companies exist to provide the services of: (1) maintaining your list on computer cards or magnetic tape and updating this list as you provide names from various sources, (2) designing and printing your promotion and keeping your promotion in storage ready for your next order, and (3) placing your promotion material into appropriate envelopes, putting labels on the envelopes, and sealing and sending the promotion to the post office.

In short, many of these companies will do almost all of your promotion work for you. Even if you are a small company, I recommend that in the beginning you use one of these mail/list houses to do some of the work for you. Some will work with very small lists, and they are exceedingly competent at what they're doing. There's a lot to learn in the mailing industry and, in order to get promotion at a low price, it often pays to employ a professional mailing house with a good reputation. It costs quite a bit to have secretaries do the work by hand. They work at one-half to one-third the speed of the mailing houses and can't compete with them. Moreover, the mailing houses have high speed equipment that is very difficult to match.

If you choose the right mailing company, you'll be able to use that company just as you would use a section of your own business. Their business depends on delivering exactly what you need at the proper time. As you might expect, however, some mailing companies are not as efficient as others, and you should be very careful in choosing the proper one.

Before you even buy a list, there are a number of questions to be answered and sets of data that you should obtain. If the mail list company is going to run your promotion operation and mail all your materials for you, then you should establish a good working relationship with them. You must know exactly what they will and will not do, as well as what they will charge you for their services. Some questions to ask list companies are:

1. As you add names to your prospect list, what method will they use to add them to their system, and what will be the charge per name?
2. Do they have a system to avoid multiple names? If you give them the same name two or three times, will they be able to avoid sending that prospect two or three pieces of promotion?
3. What type of labeling system do they use, and what would be most advantageous for your application? There are different types of labeling, some of which look very sophisticated and others of which you'd never want to send to a prospective client.
4. How long will the mailing take? You will want to know how long it will take them to send promotion out once you've delivered the promotion material to them. This delay time is very important, especially when you've got your promotion lined up with other programs within your company.
5. Will you be guaranteed that nobody else has access to your list?
6. What is the complete cost breakdown for every task that they perform for you?
7. Ask for a list of some of their clients. Call their clients and find out if they were satisfied with the mailing company's performance.

Step Three—Send Out Your Promotion Pieces

Start Now

Make a commitment to get some form of promotion out, no matter how small, and to get ready to handle the inquiries which will be coming in. You'll be able to use these promotion pieces as the foundation for your prediction of the number of inquiries. Knowing your number of inquiries, you'll be able to predict approximately how many qualified prospects you'll obtain and thereby set up an entire sales program. Without these promotion pieces, it's difficult to set up a realistic sales program which has any predictability to it.

Even if your budget is small, you can start now. Any kind of a promotion helps. Send something, even if it's only a resumé and a cover letter. In Chapter Two we discussed many different types of promotion pieces, and you can see that you can start out with some very inexpensive items.

Once these pieces of promotion have been mailed out, you will want to know how efficient the various pieces are. You'll be able to determine this by measuring the percentage of responses that you receive from using different promotion pieces. Throw out the poor promotion and continue with the types of promotion that work the best. By upgrading, you will eventually make your promotion campaign less expensive to operate, and it will become much more efficient.

PROFESSIONAL PROMOTION STEP BY STEP

This Promotion Plan Saves You Valuable Time

The purpose of having a large number of promotion pieces mailed out is to obtain an effective promotion machine within your company that will relieve you of looking for prospects. Inquiries will constantly be coming into the company, and you won't have to put your attention on getting new leads. There's no doubt that you need help getting inquiries. If you don't think you need this help, then you're sadly mistaken—you must have a large number of inquiries so that you can work on selecting qualified prospects. Furthermore, it's ridiculous for you to be spending time trying to get inquiries when it can be done using mail-type promotion and advertising. Spend your time dealing with qualified prospects, giving sales presentations, and writing proposals. Your time is very important and must be spent in dealing with clients; clients should always receive the highest priority, and thus you should not spend time looking for inquiries. The ideal situation is to have so many inquiries that you'll have to hire extra help just to handle them.

This promotion plan is based in part on a principle that prospective clients must have items which represent your company that they can hold in their hands. Verbal communication and verbal promotion is important, but it's not enough. One of the facts you're going to have to get used to is that the people who require your service are very difficult to find, and you're going to have to use some form of a "blitz" technique with promotion pieces that introduce your services. Opening the mail and seeing your name on a piece of promotion has a positive effect on technical professionals. We are actually very fortunate that technical professionals open their mail and actually look at promotion literature. It is rare to find a public that actually responds to the mail that is sent to them.

Promote to Different Groups

Break the list you created in Step Two into different categories; for example: program managers, engineers, department heads, salesmen, etc. Decide on an appropriate promotion piece for each group. In this example, program managers might receive a resumé and a cover letter, whereas engineers might receive a reprint of an article that is pertinent to their work. A department head might find a company brochure helpful. You will have to be the judge of what promotion piece is best to send to a particular group. Independent contractors who are only concerned with one specific type of contract will only find it necessary to send out a large number of resumés to managers. As your services expand, however, you should promote to more than one group.

Step Four—Handling Responses

Respond Quickly

Perhaps the most important rule to remember in this type of promotion plan is to respond very, very quickly to any type of inquiry. Try to have someone answer the phone at all times. An answering service or a phone machine will be sufficient for most technical people starting a new practice. If you have not made arrangements for your phone to be answered when you are not in the office, there will be a time delay that could be responsible for the loss of a client. Most often a client needs help on a problem now. If he's interested in having you help him, he will be interested in others as well. Don't delay in responding to him.

Even if you are very busy and you feel that you can't take on any more work, you should still respond quickly to a prospective client. Tell him that you are very busy and you are not available. If you know of someone else who could do the job, consider subcontracting him to do the work for you. In any case, respond quickly so that the prospective client can determine your eligibility. He will appreciate your quick response even if you are not available.

Under no circumstances should you ever ignore an inquiry. If you are extremely busy, have someone else answer for you. Ignoring an inquiry is a form of business suicide.

If You are Not the Right Consultant

You will get inquiries that are not for the type of work that you specialize in. Don't pass up such inquiries until you have asked yourself the following questions: Do I know someone else who could do the job? Do I know someone who could give the prospective client information on where to find more assistance? Is there a potential finder's fee or subcontract available with this prospect? Does this prospect know or work with managers who would be hot prospects for a contract with my firm?

Here is an example of how a good lead can be generated. You receive a response from a program manager in a large company, and you tell him that your specialty is not suited for his task. However, you tell him of someone you know who could do the job. He then hires this person and is satisfied with his performance. You now call the manager back and ask him to set you up with other managers who could put you to work. He'll be sure to assist you.

Handling a Hot Lead

When you get a response that looks good, be sure to qualify the prospect to see if his type of work meets your objectives. You can get a great deal of information over the phone, and you should not waste a day by making a useless trip to his company. Handling the hot prospect and determining if he is really qualified is covered in more detail in Chapter Six.

5

Subcontracting to Job Shops and Agencies

In addition to the methods of promoting yourself which I have described in the foregoing chapters, another popular way of getting contracts is to subcontract your services to a contracting firm, also called a **job shop.** There are hundreds of these firms who make their income by leasing engineers and data processing professionals to firms in need of temporary technical assistance.

There are increasing numbers of engineers and data processing professionals who are finding positions through job shopping. Because job shopping is often confused with consulting or working for a consulting firm, I'd like to clear up a few misunderstandings about job shopping, and show you how it can benefit you.

How Job Shops Work

Technical professionals who go to contracting firms for temporary assignments are called job shoppers. Most often they are employed by the job shop, which in turn leases them to the client company. Even though the job shopper is usually employed by the job shop, he very rarely, if ever, interacts with the job shop directly. In the main, the job shopper reports directly to the client company and keeps the same hours as the salaried employees of the company. Usually job shoppers are not in business for themselves, but are employees of the job shop and are paid on a weekly or biweekly basis.

In engineering and programming, job shoppers earn between twenty and fifty dollars per hour. In most cases, taxes are taken out of their incomes directly, and the job shop usually makes between 20 and 30 percent of the employee's gross income. For example, if the job shop charged their client $1400 a week for your services, they would take approximately $350 a week off the top for themselves and pay you the remaining $1050. You would not receive $1050 in a lump-sum payment; they would take out taxes, social security, etc. Some job shop companies have full company benefits such as medical and other forms of insurance, and in that case, they would also take out money from your income for that as well. In the case above where they were charging their client $1400 a week, they would not usually tell you that they were billing the client at that rate. Instead, they would advertise an assignment at $26.25 per hour which, for a 40 hour week, comes to $1050.

If the position they found for you was out-of-town, they would also add to your base pay a supplement that they refer to as a *per diem*. This *per diem* coverage is to pay for your room and board while you're away. It can range from as low as one hundred dollars a week all the way up to several hundred dollars a week. They will also pick up the travel tabs on long-distance assignments.

Although job shopping firms are located throughout the United States, they are most heavily concentrated in the major industrial cities.

The two major advantages of working for a job shop are that the pay is very high compared to most engineering or data processing positions, and that you don't have to do any marketing of your own services. One of the major disadvantages of job shopping is that you don't usually reap the benefits of being self-employed.

Job Shopping for Consultants and Independent Contractors

A very well-hidden fact is that those of us who are consultants can also get contracts through job shops. You do not have to be an employee of a job shop in order to have them find you a contract. Instead, if you can show them that you're legitimately in business for yourself and that your background is sufficient to work as a consultant, they will try to find you a job with one of their clients. They will call a company and tell them that they have a person that they would like to put in to work through their firm. They often don't even tell the company that you'll be working for them as an independent contractor or consultant. Therefore, if you contract with a job shop, you will act as a representative of the job shop. You will not act in your own behalf and will not be a representative of your own firm. The job shop company will then bill the client, let's say, for $1400 a week. They'll take about 25% off the top and they'll pay you the remaining fee, $1050 a week. If you're in business for yourself, the job shop will pay you the entire $1050, and you'll have to handle your taxes and insurance on your own.

There are some job shopping firms that will only treat you as an employee. However, over the last few years a very large number of firms have begun working with consultants who are in business for themselves. In fact, in the data processing industry, there are now job shops that would rather not hire you as an employee but would prefer that you be in business for yourself. In this case they don't have to go through the paperwork of withholding money for social security or tax purposes.

SUBCONTRACTING TO JOB SHOPS AND AGENCIES

To work as an independent contractor for a job shopping firm, you should have your own business license and your own worker's compensation insurance. You should present a letter to the job shopping firm stating that you are indeed in business for yourself and that you don't want them to treat you as an employee, but that you would rather work for them as a subcontractor.

Many of your local job shopping firms can be found in the Yellow Pages under "Temporary Employment." Another area where they can be found is in the ads in your employment section of the Sunday newspaper. These sections have ads that claim that you can make $30 to $40 an hour as an independent contractor. Call these firms and tell them that you do not want to act as an employee, but that you are in business for yourself. They'll do their best to find a contract for you on your terms.

Working Successfully with Job Shops

Don't Deal with Only One Job Shop

A characteristic of these companies is that they would rather deal with you on an exclusive basis. That is, they would rather not have your résumé go out to other job shopping firms who would compete with them in trying to find you a position. Don't go exclusive. There's nothing wrong with sending your résumé out to five, ten, or even 25 different job shopping firms. Some people send their résumé out to as many job shops as they can, and then charge a somewhat higher fee, relying on the fact that a small percentage of customers will pay the higher fee.

If you send your résumé to a job shop and they don't have a contract for you, they might tell you that the reason is that things are a little bit slow in your area. Don't believe them. If they do not succeed in finding a position for you in short order, it simply means that they do not deal with companies that have positions open that would utilize your services. There might be five or ten other job shopping agencies that could place you immediately. In the beginning, I suggest that you deal with between seven and ten different job shops. When you send them your résumé, put a note on it to have them send you back a card saying that they did indeed receive the résumé and that they are looking for a position for you. If you don't hear from them within a week, send them a second résumé and then follow that up with a phone call telling them that you want them to get a little action going for you. Call them often, and remember that these job shops are sometimes inundated with résumés and they can't give their attention to everyone. However, they will give their attention to those people who are aggressive enough to demand action.

Résumés

The way you write your résumé is very important for some job shop agencies. At the top of your résumé you should have between five and fifteen key words that are associated with the type of work you do. If you're a computer programmer, then at the top of your résumé you should list the different languages with which you're familiar, and you should also list the different machines with which you have experience. When a person in a job shop agency sees a résumé like this, they'll concentrate primarily on these key words. Also, some job shop agencies have computers in which cross-referencing is done. This computerized service of finding you a position using cross-referencing employs the key words that should be at the top of your résumé. An example of an effective résumé is provided in Chapter Two.

Give Plenty of Notice

When a job shop knows that you are available immediately, they will work a little bit harder to place you. One of the problems that job shops have is trying to find a new contract for a person who is just finishing a contract. Timing is a crucial issue. They know that if they can't place you immediately, they'll probably lose the opportunity to profit from your services.

Ideally, most shops would like to have a month's notice so they can place you in your next job; even two months' notice is sometimes needed for some positions. You should therefore do your best to contact the job shop agency when you know your present position is going to be over.

Body Shops

Working for a job shop as an employee or as an independent contractor can be financially rewarding. However, there are some potential problems in working for a job shop with which you must be familiar. Some job shops have reputations for being what are called "body shops." What they do is just provide as many bodies for potential placement as possible. They have very little regard to a client's problems, and they're not highly concerned with the quality of individuals that they send to the firm. Investigate a shop before working for it. Ask them for references. How long have they been around? How long a delay will there be between your work and your pay? Remember that when you work for a firm through a job shop, you will not be representing yourself as being self-employed. Rather, you'll be a representative of the job shop. It's true, you might be subcontracting to the job shop, but you're still going to work as one of their associates. If they have a bad reputation, or if they're in the process of earning one, you can be sure that some of it will rub off on you. Although you can get around it a little bit by doing an exceedingly good job, it won't be an ideal situation.

By-Passing the Job Shop

If you wish to work as an independent contractor for a job shop, you will most likely have to sign an agreement with the job shop that you will not contract independently with the client as a consultant or as an employee. Usually job shops make special arrangements if you want to become an employee of the firm. In this case, the firm would have to pay the job shop a finder's fee that is similar to the fee that personnel agencies are paid. If you're acting as an independent contractor and you want to take on your own contract with the company that was originally found by the job shop, you might run into trouble. The job shop would have the right to sue you for any losses that they incurred by your leaving them and going directly

SUBCONTRACTING TO JOB SHOPS AND AGENCIES

with the company. If the job shop complained to the company, the company might refuse to take on your services. If that were the situation, you would then have a job shop that was upset with you, and a potential client that did not want your services. Therefore, if you decide to work with a job shop, you should first find out what their policy is with regard to your taking in future business with their client.

Job Shops Pay Promptly

An advantage of working for a job shop is that you'll be paid promptly. There are some firms that take quite a while before they pay their consultants. Some large companies wait from 90 to 120 days before they pay their bills. If you work for a job shop, that shop is often obligated to pay you every two weeks. Therefore, if you work for a job shop, your cash flow can be more predictable than if you bill the company yourself.

Contract Termination

Another factor that you must contend with when you are working for a job shop is that you might get very short notice of your contract termination. Though some job shop contracts run for a considerable period of time, some are quite short. It's quite possible that you could work on a contract for a couple of months, and then to your surprise find that your contract was over on a week's notice, or even less. If you're an independent contractor and you have a good rapport with your own client, they'll give you a good idea of how long your consulting contract will last. Quite often, you'll know your client's budget and how much he has allocated for your contract, and you'll be able to predict when your contract will be over. When you're working for a job shop, this is not always possible. You might be at the mercy of the shop's telling you when they expect the contract to be over. In some cases they don't know themselves, and in others they don't really care very much. If this is the situation, your predictability of when your next contract should start is taken away from you. This, of course, can cost you money and make your work unpleasant. You should try to minimize these types of problems by working for a professional job shop and staying in close communication with them with regard to the status of your contract.

Job Shop Fees

Some contractors resent paying a job shop 25% of their gross income. They feel that it is too high a fee and that a firm shouldn't take so much money since they are not performing the work. I disagree with this. Taking 25% of a gross income for finding a job for somebody is not a very high fee. It is seldom easy to place somebody into a technical contract. I've been involved in finding contracts for consultants, and I have found that it takes considerable work to place someone in the proper position. In addition to the amount of legwork that must be performed by the job shop, they also must borrow considerable sums of money to ensure that their employees and subcontractors are paid promptly, and high interest rates on such borrowing increases their expenses substantially.

A Common Situation

A situation that occurs commonly when you are looking for new contracts necessitates that even if you have no interest in working for job shops, you should still understand how they operate. As an independent contractor, you might find a prospective client for whom you definitely want to do some work. He tells you that he wants you for the job, but tells you that his company does not hire independent contractors—only contractors who work through job shops. Furthermore, he tells you that there are one or two job shops that have exclusive agreements with his company. These exclusive agreements state that the job shop will provide all the temporary help that the company needs. In a situation such as this, the prospective client will want you to work through the job shop, even though he knows that the job shop will charge him more for your services than if you were contracting the job by yourself. In most situations, this does not bother the client.

Even though the client tells you that they have an exclusive agreement with a job shop, you should not agree to work for this agency immediately. In most cases the agreement that a company has with a job shop is such that the job shop is supposed to supply *contract labor type help*. Job shops are not usually expected to supply consultants and specialists. Therefore, ask your prospective client if his company hires consultants. He will most likely say that they do have agreement forms for consultants. Tell him that you do not want to come in on a contractor (job shopper) status, but that you would rather be hired as a consultant. In every situation that I have faced similar to this, I have found that the company always has special forms available for consultants that will not violate the job shop's agreement.

If they are still insistent that you work through the job shop, explain to them that you will still demand your fee, and that the job shop will charge them more money because they will add on their commission in addition to your fee. If the company agrees to this, and if you agree to work for a job shop, you should call the job shop and tell them the situation. Explain to the job shop that you are an independent contractor and that you will refuse to work in an employee status. When you're speaking to the job shop representative, remember that you are holding the cards. Basically, you are offering them a substantial profit for a situation in which they have to do very little work; all that's necessary is for them to put you on their payroll. They'll make a considerable profit from your efforts. With this in mind, don't let them try to convince you that you have to become an employee. You don't. Drive a hard bargain, and make sure that you get all the terms you want.

Documentation

If you decide to work as an independent contractor or as a consultant for a job shop, you should have the following documentation available to present to the job shop.

1. A copy of your business license. This license is obtained from your city clerk.
2. If you are incorporated, you should have your Articles of Incorporation, as filed with the Secretary of State in the state of incorporation.
3. If you are a member of a partnership, or if you are doing business under a fictitious business name, you should have a copy of the relevant fictitious name statement filed with the appropriate county agency.
4. A Certificate of Insurance showing evidence of Worker's Compensation coverage.
5. A Certificate of Insurance showing your liability coverage. The job shop might want to be named as one of the insured through your liability insurance.
6. An Employee Identification Number, which is furnished to you by the IRS.

NOTE: Be prepared to invoice the job shop on a weekly basis. A time card should be attached to the invoice. This invoice should just give the amount owed to you, and should not include the hours worked; the hours worked should appear on the time card. The weekly invoice should be submitted on a company invoice or on your company letterhead.

A Summary of Job Shopping

As I see it, the major benefit of working for a job shop is that you don't have to promote your services. The major disadvantages are (1) you aren't building your own client base, and (2) you must work for 25% less income. However, if you are just starting out, you might find that working for a job shop is an easy way to start. After you become accustomed to working on temporary assignments, you might want to start promoting yourself. My experience has been that about fifty percent of the technical professionals that I have assisted in starting their own practice have had some experience with job shops. It is a very easy way to break into consulting, and I suggest that you give it more than just a passing consideration.

6
Turning Prospects into Clients

As you can well imagine, all the promotion in the world is wasted if you treat your prospects incorrectly. The material in this chapter covers important topics such as interviewing properly, finding the client's objectives and presenting yourself in the proper fashion. If this is done correctly, the prospective client will see that you're the right person to assist him in meeting his objectives. In short, you will have landed another contract.

TALK TO THE CORRECT PERSON

Your first step is to arrange interviews with the prospective clients. You must conduct an interview with a prospect client who has the authorization to hire you as a consultant. This is a place where mistakes can be made. You must always be on your toes not to agree to an interview with someone who does not have the authority to hire you. A typical situation that can occur is as follows: An engineer calls you into a company and tells you that he is very interested in having you do some subcontracting work for his group. Later on you find out that this engineer did not have the authorization to hire you, and the person who is in charge of the funding indicates that the funds will not be available for this work for another two or three months. By falling into such a trap you can waste an entire day.

A client is not just a company. A client is an individual within a company who is responsible for reaching some specific company goals. He is a specific person, not some generality like a board of directors, and has the authority to bring you in as a consultant, direct your activities and end your contract. He might need a signature or two to get you on the program, *but it's his decision that counts*. Just because somebody asks you to come in for an interview doesn't mean that he is in charge. How do you find out? Ask!

Before you leave for an interview, ask the prospect on the phone who the person is that has the final authority to hire you as a consultant. He will either say that it is his responsibility or that it is someone else's who is higher up on the ladder. If he says that the responsibility is his, you will have to use your own judgment whether or not he is correct. If he hesitates, you should suspect that he does not have the authority that you are looking for. In this case, you can call a higher authority in the company and make sure that he understands that you're a consultant and you're coming in for an interview. If you take these precautions, you will eliminate the loss of a few hours or even a full day by visiting the wrong person.

DON'T TALK TECHNICALLY ON THE PHONE

Once you have determined that you're speaking to the right person, the first thing to do is to find out what he needs from a consultant. When he states his problem, make sure that you do not get into any form of a detailed conversation that shows him you know how to solve the problem. Just say that you have experience in the area of his problem and that you can help him. Tell him that you would like to make an appointment with him to discuss the matter in more detail, and that a meeting would provide both of you the opportunity to discuss the problem in greater depth. The purpose of an initial phone conversation is to *determine if you are interested in a contract* and to *set up an appointment for an interview.* Trying to solve technical problems over the phone or even discussing them at this point is incorrect.

THE PRELIMINARY INTERVIEW

The First Few Minutes

The objective of this step is to spend a short period of time with the qualified prospect to make him feel at home in your presence. Your objective is to set a foundation for a relationship that is based on trust. The last thing that you should do is walk into his office, open your briefcase, and immediately start speaking about business and technical problems. Upon entering his office, look around for evidence of his interests. For example, if there is a golf trophy on his desk, you might start a brief conversation about the condition of the local golf courses. This will give you some common ground on which to begin your acquaintance. If he is a technical person and there is no evidence of personal interests in his office, you can always glance at his bookcase to determine the subjects that interest him. Find out something about him. How long has he worked at this firm? What type of an educational background does he have? How does he fit into the project on which he is working? If he is interested, you might tell him a little bit about yourself and what it's like to be in business for yourself. Gradually you can enter into a semi-technical conversation and tell him about the kind of work with which you've been involved over the past few years. If he is at all talkative, you should spend the majority of your time listening to him. Being a good listener is always important, and in a situation such as this it will help him relax in your presence. During this part of the interview, make sure you don't get into any deep business or technical discussions. Just do whatever is necessary to make the atmosphere relaxed and congenial. Although this initial meeting period only lasts a few minutes, it will set a positive tone to the rest of the interview.

What Are the Client's Objectives?

After you have developed a good rapport with your prospective client, you can start speaking about business. More accurately, you can start listening. *At this point in the conversation you are not trying to sell your services.*

You must first determine some very important facts. First, you must find out what is really needed and wanted by him. You must find the specific problems he wants you to solve. Determine how these problems fit into the overall picture and then find out how solving these problems will help him meet his personal objectives with respect to the project. A prospective client will usually decide to give you a contract if he sees that you can help him meet his personal objectives. For example, a prospective client might be totally overloaded with work, and at first he tells you that he'd like you to write a particular type of computer program. Upon further investigation, you find that his personal objective is to take a three-week vacation. After telling him that you can write the program and that you will do your best to make sure that he will get his vacation, you can be sure that he'll be interested in giving you the contract.

You might also ask him how he fits into the overall picture. How do his responsibilities fit into the project and the company's objectives? Most prospective clients feel their work is very important. Most often they'll be happy to tell you how they fit in and how their role is necessary to the success of the project.

Don't Be Afraid to Show Your Ignorance

While conducting an interview, some people aren't willing to show their ignorance, and consequently don't really understand what the client needs and wants. This is very foolish. It's extremely important that you understand what the real problem is before you agree to take on a contract. In my interviews, I have always asked extremely basic questions to guarantee that I really understand the client's problem. I have never run into a situation where the client considered my questions so basic that it might be a sign of ineptitude. In fact, many prospective clients have appreciated my candor, and I'm sure you'll experience the same thing. It often takes a considerable amount of time before you really understand *what the problem is* that has to be solved. Only if you understand what the objectives and real problems are can you make an accurate estimate of how long it will take you to do the job. Therefore, during the interview, feel free to ask as many questions as you like.

Discuss Your Pertinent Experience

After he has explained the tasks that he needs accomplished, you should then show him how your background relates to solving his problems. You should not go into a discussion of any experience that you have had that does not apply to his specific problem. He is only concerned with whether or not you have the experience to do his particular job. Therefore, only show him the factors in your background that will help him meet his objectives.

Even if the interview is going exceedingly well at this point, you should make sure that you do not decide whether you want the job at this point. Just because you know what technical task has to be accomplished does not mean you should jump the gun and make an early decision. There is much more data to collect.

Assess the Client's Viability and Business Ethics

At this point it is necessary to collect some very simple facts that will assist you in deciding whether you should take a contract with this prospective client. It is very important that you become involved with a company that is expanding. A company that is contracting is filled with political problems, and for this reason you should stay away from this type of firm. Simply ask what their sales volume was like over the past year or two and find out how the company is presently doing. Also, find out if the department in which you will be working is expanding or contracting. Again, you don't want to be involved with losers. If you work with an expanding department, you can be assured of receiving considerable follow-up business if you do a good job on your first contract.

You might also ask why they need help and why they are considering an independent contractor for the job. Is it because they don't have anybody else to do it, or have they just laid off a large number of people who could have done the work? What you're trying to find out with questions such as these is whether or not they are having major internal problems that could interfere with your working on an interesting and profitable project. You should not just be looking for a contract. Rather, you should be seeking interesting work in a pleasant environment that leads to both profit and enjoyment. Also, your success will depend on follow-up contracts, and these can be obtained only from companies that are doing well and expanding.

If you have any doubts about their financial viability, you should determine whether or not they are willing and able to pay you. You can simply ask if they are in a financial position where you might have trouble getting paid. The answer to this question is usually that they do have enough money to pay you; after all, you can't expect them to say that they're not planning on paying you. Even though they will give you a "yes" for an answer, notice whether they are hesitant in answering your question. If you notice any hesitation, you should continue with your financial investigation. If it is a small company, ask how long it will take you to get paid after you invoice them. If a small firm tells you that the time is between 90 or 120 days, then you should be concerned that this company might be in sufficient financial trouble that you may never get reimbursed for your efforts.

With a large, fairly stable company, it is not uncommon in these times to have to wait three, four, or even six months for your paycheck. With inflation and the cost of money being as high as it is, many companies will do everything in their power to pay you as late as possible. In a similar fashion, you should do everything in your power to make sure that you get paid promptly.

Don't go on the assumption that just because a client has the money that he will pay you. He could be rolling in dough and be a crook. If you doubt his integrity, get the names of other consultants and companies that he has contracted with and give them a call. You might be surprised to find that underneath that friendly smile he's not so lily white. When it comes to getting paid, decide to get the facts early to avoid getting burned later. Take my word for all this and don't repeat my mistakes. Ask!

Remember that at this point you are not trying to negotiate with your prospective client regarding how or when you're supposed to get paid. All you're doing at this step is collecting enough facts so you can make a rational decision as to whether or not you want to work with him.

Another factor which is very important in taking a close look at your prospective client is to determine what kind of business ethics his company follows. One of the best ways to determine if their business ethics are admirable is to examine the quality of products they produce. If they produce high-quality products, that is a good indication that the company is guided by ethical people. However, if they're producing shoddy products, then you can expect that their business ethics leave something to be desired. Don't let the simplicity of this fool you; it's an extremely accurate indicator. They depend for their survival on their customers. If they're willing to deliver an inferior product to the customers that they depend on, you can be fairly certain that they will treat you in a similar fashion.

To Whom Will You Be Delivering Your Service?

If the company has a good reputation for high-quality products and service, still more data must be collected before you should decide to take a contract with them. Next, find out who will be your principal contact in the firm. Will you deliver your services to the person interviewing you, or are you to report to someone else? If you are to report to someone else, ask to meet that person. You will be spending a considerable time with this individual, and you should find out if you will be working under pleasant conditions. You should make sure that this person is rational and will not create any major problems for you on your contract. I once took a contract with a client whom I respected and whose company I enjoyed. Unfortunately, part of my assignment was to work with one of his subordinates who at times went into irrational tirades and verbally attacked everyone around him. I would have turned down this contract if I had carefully investigated with whom I was to work on the project.

I must tell you about another experience which I had early in my consulting career where I made a mistake that led to the loss of about $1,000. This is the only time that I did not get paid for a consulting contract. I was asked by an acquaintance who owned a small engineering firm to work for him for a week and help him write a proposal. I had known him for a year, and I didn't give much thought to his business ethics. This was a mistake. I learned the hard way that just because you happen to know the person who hires you is no guarantee that you'll get paid. This fellow seemed to get along with everyone and gave the appearance of being very friendly, but in reality he was thoroughly insincere. It turned out that he was in debt to a large number of people and that shortly after I worked for him his company went broke. I was not the only person who got burned by being associated with him. My mistake was that I did not take a close look at his actual production. If I had, I would have seen that he had a history of making big promises to customers and then delivering little or nothing at all.

The lesson to be learned here is a very simple one. If you're not sure whether a prospective client is telling you the truth, you can and should obtain more information by investigating his credit rating and by determining whether or not he delivers what he promises. Investigating his credit rating can be done through a credit agency or by calling a list of his previous customers, subcontractors, and acquaintances.

DECIDE WHETHER YOU WANT HIM AS A CLIENT

To make a correct decision at this point, you have to have your personal and your business objectives well defined. Your decision-making policy with regard to taking on new contracts should be laid down in advance so you can make the proper choices. If you find that the company delivers a good product and you feel that you could work well with them, then you should indicate to your prospective client that you have an interest in taking on a contract. If you still have doubts, you should collect more information and politely tell him that you're not in a position to make a decision at this time. You should tell him that you need more information and that you want to look into the project and its associated problems in a little bit more depth.

Notice that nothing has been said up to this point about his wanting to hire you as a consultant. Nor have we indicated that this is the right time for you to start selling your service or presenting yourself in such a way that you want to convince him that you're the right person for the job. That is because *prior to closing him on yourself, you should first decide whether or not you want to work with him.* Only after you have made the decision to go with him should you attempt to close him. If you try to close him before you have decided that you want the contract, you might find yourself working on a program in which you're not interested.

If you decide not to take the contract, you should end your discussions positively so that you will have the option for future business. Explain to him that this is not the type of contract for which you are looking, but that you might be able to do business with him in the future.

THE CLOSE

If you have decided that you want him as a client, your next step is to show him how you will employ your experience to help him meet his objectives. At this point you do not enter into any form of negotiation. Just sell him on the fact that you are the individual who will help him meet his objectives. Stress the benefits and advantages of working with you. Sell yourself and the idea that you are the person who can get the job done efficiently and effectively.

If he wants to talk about fees at this point, you can tell him what your fees are, but don't get into a discussion of whether or not they are too high. For example, he might say to you, "I think $65 an hour is a little high." You can then say to him, "Let's first decide whether or not I'm the right person for the job. I'm sure once that's decided we can agree on a fair rate of exchange. I can do the work and I deliver what is promised. If you're interested in me, we can take it from there and negotiate the proper fees for getting the job done." *It's imperative that you keep the discussion away from money or any other details before he's decided on whether or not he wants you to do the job.* Once he has decided that you are the right person to do the work, you should then prepare to do some planning and negotiating.

If he thinks it might be less expensive to hire another salaried employee, you can show him that it will be to his benefit to hire you as a consultant. Here is some ammunition to help you in this regard. He saves thousands of dollars per year by hiring you and not attaching overhead to his budget. Due to poor management, many large corporations have raised their overhead multiplier to ridiculously high figures. What's an overhead multiplier? It's the amount by which a program manager multiplies an engineer's salary to determine how much that engineer costs his project. For example, if an engineer's salary is $35,000 per year and the company has an overhead multiplier of 2.4, then he will cost a program manager's project $84,000 per year.

Let's say that as a consultant you approach this program manager and offer your service for $1,400 per week, or $70,000 per year. By working with you he will receive many benefits. (1) The program manager can charge his program this fee with no overhead attached. Your fee goes on his books just as if you were a subcontractor or a piece of equipment. (2) He gets you immediately. He doesn't have to go through the long process of hiring an employee. (3) He charges you directly to his programs and is thus assured that you won't be ripped off by some other part of the company. (4) He doesn't have to worry about laying off his salaried employees. (5) He has happy salaried employees who think they will be laid off last (not always true) if trouble comes.

Many program managers have not hired consultants, and they are not aware of these obvious benefits. A brief reminder of these benefits to a prospective client will help you handle any misgivings that he might have about hiring a consultant. But keep in mind that in most technical situations you do not need a formal sales pitch. In most cases, all you have to do is show the prospective client how you can help him meet his objectives.

USE YOUR PROMOTIONAL MATERIALS

During the interview you should make full use of all the promotional materials that you have prepared. In particular, you should show your prospective client your notebook that contains your letters of recommendation. Also make sure that you have your list of references and any other supplementary material with you that might be needed during the interview. For example, if you have reprints of any articles that you have written, you should take them along. If the discussion relates to a similar topic, then you will be able to pull out the proper reprint and show the prospective client that you have experience in that particular area.

It is also helpful to bring three or four resumés with you to the interview. You will often interview with more than one person, and you will want to have resumés available in case they do not have copies. Remember that only those points on your resumé should be stressed that show that you have the experience to handle the particular job that they have in mind for you. Most prospective clients are not interested in experience which you have that is not related to their objectives.

A summary of past contracts is also helpful in an interview. If one of your past accomplishments is similar to his present problems and objectives, he will be interested in having a copy to attach to your resumé.

7
Giving Effective Presentations

Giving effective presentations is a small but important aspect of consulting. In many types of independent contracting, an engineer or programmer rarely has to speak in front of a group. In fact, some contractors never have to give presentations. Nevertheless, you should be prepared to speak in front of a group at all times, because it is often during these critical moments that a contract can be won or lost.

Another reason for developing your presentation skills is to improve your communication while on the job. Giving a presentation is often the most efficient way to communicate your results and recommendations to your client and his staff. There is no sense in spoiling many months of good work with an ineffective presentation.

PRESENTATION DELIVERY TIPS

Stand Up Straight

Hiding behind your glasses or having poor posture is no way to give a presentation. Stand erect and stay that way throughout the delivery. It doesn't mean you have to be rigid, but you shouldn't hunch over and hide. After all, you as the speaker are in control, and your posture should help to communicate your self-confidence.

Look at Your Audience

You're not speaking to just a group; you're speaking to individuals. If you speak to an individual with your back turned or your head to the side, you lower your communication effectiveness considerably. The same holds with an audience. When you're using visual aids, it's common to turn away from the audience for a moment to show a graph or give an example, and it's easy to remain in that position for too long a time. If you must turn around, do so briefly.

Watch Your Audience

Be concerned with whether or not your audience has their attention on what you're saying. If their attention drifts, spot this as soon as possible. Though this can usually be detected by simply noticing whether or not they are looking at you, this is not always the case, and so you must stay alert.

If he loses his audience, the good speaker does not panic or become embarrassed. Simply get some feedback from the audience as to what is wrong. There is no law that you have to give an hour's dissertation before getting any feedback from the audience. As soon as there is some serious drift on the part of the audience, question the audience and get the matter corrected. Don't let them drift. If you lose their attention, they will not get your messages. One major aspect in doing technical presentations is getting the audience to understand the logical sequence of your data. When your audience's attention wanes, even temporarily, major parts of the logical sequence of your presentation are not received, and the audience will miss the main point. Even if their attention only drifts for 10–15% of the time, all of the message could be lost.

Be Aware of Background Noise

Often in a noisy factory or in the city, the background noise can be excessive. Sometimes it reaches the point that people will not be able to hear you. Even if they can hear most of what you're saying, there will be occasional bursts which drown out what you have to say at the moment. Be aware of these noises, and if they occur, be responsible for seeing that no one in your audience misses what you are saying.

Know Who's in Your Audience

Before you give a presentation, you should make a few telephone calls and find out who'll be attending. If you aren't acquainted with the important attendees, find out about their positions and their backgrounds. If you find that a considerable number of attendees will be administrators and decision makers without a strong technical background, cut your talk back accordingly and make sure you don't overwhelm them with technical terminology. People in high administrative positions will appreciate your ability to communicate the essentials of the technical aspects without going into extreme detail.

Minimize Technical Vocabulary

Even when speaking to other experts, you should minimize the number of highly technical terms. Some speakers try to look sharp by throwing out sophisticated terminology. This is a disaster for both the speaker and the audience. Keep it as simple as possible.

Speaking to Your Prospects

A young attorney asked an old successful backwoods lawyer what his formula for success with the jury was. The old man replied:

"First, I tell 'em what I'm gonna tell 'em; then I tell 'em; then I tell what I told 'em."
"I guess that drives it home to them," replied the youngster.
"Nope," said the old timer, "Only to the ones that was list'nin."

There is a tendency for people to believe that when they speak to an audience, the audience understands what they've heard. In general, this is not the case. People have a rather difficult time understanding others even in one-to-one conversations, and the situation of giving presentations to more than one person is much more serious. When you're giving a presentation, you're hindered by the fact that you don't have as much feedback from the person or persons to whom you're speaking and, therefore, it's difficult to assess just how much was understood.

This holds true in technical presentations as well as in other types of public speaking. In technical presentations, you're very often using a vocabulary and terminology with which you might be very familiar, but which your audience might not thoroughly understand. Again, detailed and excessive technical vocabulary should be minimized.

When giving sales presentations, the people who you really want to communicate with are often managers who are not technical experts. In this case, your technical expertise could be the very thing that ruins your speech. Therefore, you must translate what you understand in your language and express it in their language. One of the most effective ways to do this is to go over your presentation with non-technical people before you give your presentation to your customer. You'll be amazed at what you'll find. Facts that you consider obvious are not obvious to someone in a different discipline.

Repeat Important Points and Get Audience Feedback

There's concern among many people to be efficient and to say things only once. I don't know where they get such an idea, but it certainly doesn't work. People seldom understand something thoroughly the first time it's stated. Sometimes they might get a general idea of your main point, but often they just don't understand it totally the first time it's stated. *Therefore, say it again.* If, after saying it a second time, you're not absolutely sure that they understood it, get some feedback from the group. If you have to, ask somebody in the audience a question that's related to what you have said or open your presentation up to questions. Get feedback and find out if they understood your critical points. Also, in stressing some of the essential themes to your message, give illustrative examples which will clarify the central idea.

If you interrupt your presentation every five minutes or so and ask them if there's anything that they don't understand, you'll be surprised at the response you will get. People often hold back questions, but when asked if they have any, will volunteer all kinds of confusions.

You should also make it a rule to pace your talk to accommodate a question and answer period. This should be kept brief and not be permitted to turn into a long, drawn-out marathon which takes you off your main point. You will have to have the discipline to give short answers and explain that you'll give more detailed information after the talk. If you know your material, you should have no problem giving a short answer.

The value of having a few question and answer periods during your talk can't be overestimated. You immediately get some feedback with regard to their level of understanding, and you can adjust the technical level of your presentation accordingly. You also have an opportunity to get them to participate, and you will find out what they consider important. You can emphasize your main points while answering a question, and thus satisfy two objectives—getting their question answered, and emphasizing one of the more important points of the main presentation. Sometimes it's necessary to show your potential customer that you really know what you're talking about. Being able to answer questions as they come up in a presentation is a very effective way of achieving that objective.

What should be done if somebody asks you a question and you don't know the answer? Some people can get pretty complicated about handling this situation. As far as I'm concerned, there is only one acceptable way to handle such a situation—be honest about it and say you don't know. Be specific about what you say you don't know. If you do know something about it, then you can tell them what you do know, but be specific and say you don't know the exact answer to the particular question. Some people believe that since they are consultants, they are supposed to know everything, and therefore think that they should not reveal their ignorance. However, it's been my experience that it very often puts a potential customer or client at ease when you reveal what you don't know. After all, there's plenty which you don't know, no matter how specialized you are. You're going to be ignorant about a large variety of topics even within your own field. So admit it. Then, get on to doing your job.

If someone asks you a question that has many parts within it, mentally break the question down while it is being asked, and answer one part at a time. If you're asked a question and you don't understand what the question is, politely tell the person that you didn't quite get it all and ask him to repeat it.

Be in Charge and Don't Go Off on Tangents

Since you should use question and answer periods during your talk, you will have to make up your mind very early who is going to run the talk— you or your audience. Often, you'll get people in your audience who will run off on a tangent with their questions and continually take your talk off course. If this is the case, you'll have to be very polite and very firm, and go right back to your main line. Under no circumstances should you let one individual control the direction or content of your presentation.

Stick to the main body of your talk. No matter how interesting things might get, don't go off and talk about anything that's not directly related to your main objectives. Remember, the purpose of your talk is to deliver and support the main points. Don't do anything that does not fulfill this purpose. Relate any and all questions back to your objectives. You might be asked a question which is very easy for you to understand. However, the understanding of the particular question is not sufficient. You must understand how the particular question relates to his (and your own) objectives. This point is especially important when you get a question from one of the key members of your audience. If he's asking a question, he's trying to clear up something that is related to his meeting a particular objective. If you don't understand what the objective is, you might answer his question but communicate the wrong message to him. By knowing your audience's objectives you will be able to determine the best way to answer their questions.

Use Illustrative Examples

Sometimes, driving your points home with pure logic is a little bit too much for your audience to handle. By bringing in some short stories, humor or illustrative examples, you can change the pace of your communication and still get your points across effectively. Very few people learn by digesting only facts. They have to see how the facts are applied in practical situations.

Avoid Memorization or Reading from a Prepared Text

There aren't many things that will put an audience to sleep more quickly than someone who has memorized a bunch of lines and then is standing before them spitting out dry facts. There's no need to memorize anything in a technical presentation. You might be dealing with facts, figures, and graphs, but they are there just to support particular features of your argument, and to amplify particular advantages or disadvantages. It's the benefits that you want to cover—benefits that your customer or client will obtain by going along with your recommended techniques. If there is a set of numbers or facts which you need to carry along with you, put them on a piece of paper and bring them up front with you. If you need them, pick up the piece of paper, and then give the audience the information. Don't feel obligated to memorize anything.

Unless you're in a position where you must quote somebody, avoid reading during a presentation. People respond to live communication. The audience is happier when they feel that the speaker cares not only about what he's saying, but also about the audience to whom he's speaking. If you're going to read to them, why not just hand out the text and let them read it themselves? When you're giving a presentation, your communication capability, as well as your technical ability, are being evaluated by your potential clients.

Rehearse Your Presentation

If you feel slightly uncomfortable before your presentation, you might want to present it to someone else first. There are many advantages to this. First of all, if you give it to people who have backgrounds similar to the customer, they will be able to give you relevant feedback. You'll find out whether you're getting the major points across correctly. It's also very helpful in practicing your timing. When giving your practice presentation, be sure to have a question and answer period similar to the one that you will have when you get together with your customer. It will also be beneficial if someone in your test audience keeps a time record of how long you spend on various topics, and how long the various questions and answers took. Also, have someone jot down the places in which you strayed and did not stick to the main topic.

Speak with Authority

The ideal speaker should speak with authority. However, if there are particular topics that you're not sure about and you don't feel you can speak on with authority, then don't fake it. Either brush up on the material and deliver it in a forthright fashion, or explain to your audience that you're not an authority in that particular topic. You don't have to look like you're an authority in everything. No one expects you to be.

Minimize Arm-Waving and Physical Gestures

Some so-called experts in speaking advocate the use of excessive hand motion, arm-waving and other kinds of "body language" to get the point across. They stress that such motion keeps the attention of the audience. Well, it does keep their attention. But it keeps it primarily on your hands and waving arms. If your objective is to have their attention go to your waving arms and not to what you have to say, I suggest you wave your arms like the dickens. However, if your objective is to have them put their attention on what you have to say, then you should minimize the body motion and just get up front and be yourself.

Pause Before Starting Your Presentation

Often in technical presentations, three or four presentations will be made in sequence. If you are to start after another presentation has just been completed, walk up to the front of the group and prepare your material. You might find that the room is somewhat noisy. If you wait a few moments, you'll notice that people will stop speaking. You can just stand there and wait politely. You may even have a few quiet words to say to somebody in the front row while you're waiting. Then, after the noise has died down somewhat, you can begin. Before you start, make sure you've been introduced. If you haven't been introduced, simply look at your audience, say hello, and introduce yourself. Remember, you're not speaking to a crowd;

you're really speaking to an individual. There may be many individuals there, but there's no such thing as speaking to a crowd. Your message is received by one person in any communication.

HOW TO GIVE AN EFFECTIVE SALES PRESENTATION

Your objective of the sales presentation is to have the potential client recognize the difference between your approach and competitive approaches. He must understand that he will receive the most benefits and least disadvantages by going with your approach. If he understands this, he's sure to deal with you as the consultant for the job.

One of the most important factors of an effective sales presentation is knowing who the important people and decision makers in the audience are. Whether you're giving your sales presentation to one person or to an entire audience, you still must know who the decision makers are. You must then direct your communication to those individuals and make sure that their questions are answered. You must know who they are, what positions they hold, and what the level of their technical understanding is.

Very often, managers will be holding the decision-making positions, and it is a rare manager who has a high degree of technical competence as well as administrative ability. If you start speaking in technical language that they don't understand, you're doing worse than giving no presentation at all. You'd have a better chance staying home and hoping your competitors fall on their faces so that the potential client calls you in to take over. Don't fear to ask what the decision-maker's background is and what level of technical experience he's had. You should speak with him in private, and you should definitely find out what he needs to know. Just because you have stated the correct solution does not mean that the decision maker understood it. After all, that's what he's after—a little bit of understanding. If he has any confusions about what you have presented, you're headed towards a misunderstanding and no contract.

Before you start your presentation, state the problems that you'll be addressing in writing or on a viewgraph, and show your audience that you know exactly what they need and want. This information, of course, was obtained from them originally, but you'll be showing them that you have understood what they've communicated to you with regard to what they need. When you state the problem, look to see if there is full agreement. If there is any disagreement, inquire and find out if there is any information you haven't been given.

Your next step is to state very clearly that there exists a variety of approaches to the problem and that there are a few solutions. Indicate that you have investigated the possible solutions, and that you've made a trade-off analysis and have arrived at a solution that is most beneficial to their situation.

Give a brief description of each approach, and identify features, advantages and disadvantages of each. For each advantage, indicate what the benefit is. For each disadvantage, indicate what the customer might lose.

After this is done, show how the trade-off was performed and how you arrived at your decision on the method of approach.

Scheduling and Costing

Present Scheduling Estimates

When your potential customer has a firm idea of how you decided on the approach that you've taken, you're now prepared to show him how you've prepared your scheduling and costing estimates. I want to stress that the scheduling and costing estimate should be done after the customer has a firm idea of how you came to your conclusions. Preferably, he is already sold on your technique. If he's sold on your technique, he's going to be very interested in how the scheduling for the project should be done and what the estimates for the financial side of the project will be. It can happen that scheduling and cost estimates are given to a potential customer who's not really sold on your proposal. In this case, it is very difficult for him to put his attention into the scheduling and costing of something that he's not quite sure is feasible.

Present a Costing Summary

Two costing sheets should be given. One is a summary sheet which gives all the costs spread over time. And the second is a detailed costing report showing that you've gone into great detail estimating the manhours and materials. Don't emphasize the detailed report. Show him that it's there, and let them recognize that you've done your homework. Your objective with both the scheduling and costing reports is not to show that you're exactly right. After all, these are only estimates. Your objective is to show that you're thorough.

Don't Ignore Their Schedule Problems

In technical consulting, you will be dealing with clients who have programs and schedules of their own. One of the things you might find very helpful is to obtain the schedule that they are working on and put it into graphic form. Then, after you've presented your schedule, place their schedule next to yours (or if you are using a viewgraph machine, place an overlay of their schedule on top of your schedule) and show them that the schedule you've presented fits exactly into theirs. Show that there exists a great deal of coordination between your project and their program. By demonstrating that you haven't ignored their schedule, you'll show them that you won't go off on a tangent and that you will meet your delivery date.

Stress Your Experience

After answering questions, you should emphasize your experience; this has a strong effect on any potential customer. Give him a listing of all similar projects that you have worked on and demonstrate that you can do the job. *Before ending the presentation, you must stress the fact that you're experienced in this kind of work, that you've proven yourself before, and that you deliver what you promise.*

Audio Visual Aids and "Feelies"

Two companies were competing for a contract, the winner of which would be hired to design a particular bridge. The first company came on and gave a two-hour mathematical presentation. The second company gave a presentation which included three large-scale model bridges, and demonstrated the weakness in two of them. The people in the audience were invited to weaken some of the members in two of the bridges. The scale-model bridges collapsed right in front of the audience. When they went to weaken similar members in the third bridge, the bridge didn't even budge. After the presentation, the consulting company rebuilt the model bridges and brought them back to the customer to demonstrate again how the weakening of certain members in a certain type of design could ruin the entire bridge. It was the second company that got the point across. They won the contract.

The spoken word is limited, and when it comes time to transmitting abstract ideas, the spoken word rarely does the job at all. I received a good deal of my training in physics and mathematics, and I obtained a reputation of being able to solve abstract problems. I could often employ mathematical techniques that others were having trouble with. Thus, you'd think that a person with my training would be able to understand mathematical and other abstract-type presentations. I'll let you in on something. I have not fully understood most of the abstract presentations I have seen. The ones I did understand I had seen before, or had enough familiarity with, so that I could recognize what the speaker was talking about. Now, if dealing with abstractions was part of my specialty, and I had a hard time understanding many abstract arguments, where do you think the average listener stands whose specialty lies in other areas besides abstractions? Let me clue you in. The average listener's awareness in technical meetings is so low that he doesn't even know that he's lost.

When you're giving a talk, you need absolutely every bit of assistance you can drum up: audio aids, visual aids, and anything you can find to have your audience get their hands on. That's right—"feelies." These different types of aids help you accomplish two very important purposes. First, they help the listeners participate more in your presentation. This tends to remove the serious and academic atmosphere and promotes a more relaxed informal mood. Second, these aids make abstract concepts real and make difficult ideas simpler.

The most effective aids are three-dimensional models and objects that your customer can hold in his hand. The ideal visual aid would be a model which could be taken apart and broken down into component parts. This is the most powerful type of demonstration, and it's the one that's used the least.

The next most effective types are two-dimensional aids, which include engineering graphs, charts, maps, schematics, photographs, and even an occasional cartoon. The two-dimensional type aids would also include blackboard drawings, slides, overhead projector (viewgraphs) techniques, and motion pictures.

Home television cameras and video tape recording setups are now available at extremely low rates. You can rent them or, if you're going to use them often, purchase them. If you're in any kind of a field where you'd like to include the audio as well as the visual communication, then the video tape recordings are excellent.

Visual aids can also be used in a destructive fashion. Many talks have been ruined due to the fact that vast amounts of incomprehensible information were displayed on viewgraphs or slides. They were shown to an audience who didn't understand them. Very often, complex mathematical equations are shown on viewgraphs with the incorrect assumption that the audience is aware of what's going on. Some people present complex ideas on viewgraphs in order to look sophisticated. Trying to outsmart the audience in such a fashion is ridiculous. It doesn't work, and the person who tries such a maneuver loses very quickly.

Professional Presentations with an Overhead Projector

Small consulting firms usually don't have access to a graphics department which can prepare professional slide material. Consulting firms are often faced with the problem of sending material to be made into slides to a graphics shop, only to find out that there's a significant time delay, and the graphics shop cannot prepare the material in time for a presentation. However, if an overhead projector is employed in a technical presentation, the preparation can be accomplished within your office with the techniques described below.

The advantages of the overhead projector are many. The chief advantages regarding technical consulting are as follows:

1. Transparencies of graphics and technical material can be made by use of a blank transparency and the original placed in a photocopy machine. The blank plastic is placed in the blank paper compartment, and the result is an excellent transparency.
2. The transparency can be changed while you're giving the presentation by adding extra information by writing with a grease pen. If new data is obtained prior to or during the presentation, the transparencies can be altered immediately without a degrading effect on the presentation.
3. Transparencies which must show a great deal of detail can be constructed. Even multi-colored transparencies can be used to show the intricate parts of complex machinery.
4. A technical presentation can be coordinated by presenting written words on the transparencies along with the speaker's remarks to emphasize the important points. Sometimes placing one phrase or sentence on a transparency to get the point across can be very effective.
5. Two or three transparencies can be overlayed to make comparisons of various data. Often, it is desirable to compare two graphs, and by preparing a simple overlay the information can be transmitted immediately.
6. You don't have to use a long pointer which distracts the audience.

Instead, a sharp pencil point will create a shadow on the screen, and you can indicate exactly what you wish to emphasize without waving your arms.
7. It is not necessary to turn your back to the audience while giving a presentation with transparencies.
8. The overhead projector system is exceedingly simple and inexpensive.

If you're using your own projector, always carry an extra projector bulb along with you. Sometimes there's a place within the projector where you can tape a projector bulb (which is still in its original box) to the inside of the projector. If your bulb burns out, you'll always have a spare available.

Don't Oversell

There comes a time in every successful presentation when you recognize that your client is sold on you. At this point there is a rule you must follow and must not violate. *Never continue your sales pitch beyond the point where your customer is sold.* Anything you say from that point on only adds to the possibility that he'll change his mind.

In summary, excellence at the art of delivering presentations is attained only through knowing your audience and what their level of understanding is, and then delivering your presentation to them at that level. There is no substitute for developing your ability to observe your audience. And, like every other skill worth attaining in life, this requires practice.

8

Determining and Structuring Your Fees

How much should you charge for your services? Should you undercut your competition? What will a prospective client think when you tell him your fee?

When you go into business for yourself, you perform many duties within your company. You are still a technical professional, but you're also a businessman with the determination to show a healthy profit, and if you deliver a high-quality service, you deserve one. This chapter shows you how to choose the proper fee structure and avoid charging too low a fee—a sin I have seen practiced by too many consultants. As we shall see, choosing the proper fee structure is of crucial importance, and correct decisions in these matters depend on many factors.

DETERMINING YOUR FEE

To get an idea of the proper range for your fees, the first thing you should do is to get in touch with a few program managers. Tell them that you are planning on going into consulting and that you would like to know the range of rates for various types of consultants. You can explain to them that you do not need exact dollars and cents figures; estimates are close enough. Program managers will be happy to tell you the various ranges that they have seen. After you speak to five or six program managers, you'll get a very good feel for the range of fees.

There exists a simple formula for determining what an approximate per diem rate would be for general consulting. Determine the amount of money you would make as an employee per day, and multiply that daily salary by three. That figure will take care of your business expenses, if you have a reasonable amount of business activity over the course of the year.

Another method is to send your résumé to five or six job shops and ask them what the going rate is for someone with your background. Take the rate they give you and increase it by about 30%, and you will have a proper range for your fee.

As a standard practice, large consulting companies charge three or four times the rates paid to direct salaried employees. You should make a list of the large consulting firms in your field and determine their rates. Since you, as a smaller firm, have less overhead and promotion cost, you will find that you will be able to undercut their fees considerably, and still charge a much higher fee than you obtained as an employee.

Some Approximate Figures

Although the figures below should not be used for all situations, they will give you an idea of the approximate fees being charged in 1982. The actual fees that are charged will depend on the demand that clients have for your work, as well as your level of expertise.

Senior Scientist or Project Engineer	$500 per day
Senior Engineer	$400 per day
Engineer	$300 per day
Programmer	$200–$300 per day

Recognize That You Are Unique

Sometimes it is difficult for a consultant to recognize just how important and unique his services are. Often he could be charging much higher fees and still have satisfied customers. Most specialists undercharge. Don't fall into this trap.

If you deliver an excellent service, you should realize that compared to most employees, you are a very rare breed. Though in general there are few people who understand this, it is true that the ones who count, i.e., managers who really know their business, *do* understand this. How do you put a price on quality? Customers will pay large amounts of money to deal with a consultant that delivers what he promises. Therefore, determine what it takes to deliver an excellent service and *charge that amount.* Under no circumstances should you put all your attention on what your competition is charging; especially when your competition is not doing as good a job as you. *Do a damn good job and charge whatever you can get for it.* Remember, your ultimate objective is a satisfied customer who will use you again and is interested in your success and survival as well as his own.

If someone looks at the cost of your services and questions the price, don't get pulled into answering any generalities such as "Why does it cost so much?" or "That's a lot of money." Your response can be, "Compared to what?" Ask them specifically what they are making their judgment on. If they tell you that some other company offers similar products or services, pull out your brochures and show them specifically what the differences are. Then back this up with testimonials from satisfied customers.

DETERMINING YOUR FEE

Are You a Specialist?

The general rule that can be used as a guideline in engineering and data processing consulting is that pricing becomes less important and less fixed as services become more specialized. If you're a specialist and you have no means of comparing your services to others, you should charge a very high fee. Consultants who are specialists know that because they have information that is unique to them, they can charge extremely high rates. There is nothing wrong with this.

Of course, minor problems can occur when you're a specialist with high rates and a client does not understand the value of your services. Therefore, there's a possibility for a misunderstanding right from the beginning when a client, even if he is a well-educated and competent businessman, does not understand what you have to offer. Often a specialist is called into a project because he's been recommended by other technical people. When a prospective client faces a fee of $700 a day, or even $1000 a day, he can become rather surprised and not have any idea of what to do or say. As in most cases, it's your responsibility as a consultant to educate your prospective client in this case as to what you have to offer. One of the ways which you can emphasize the value of what you have to offer is to spend an hour or two with the technical managers and the business executives explaining exactly what you do, the type of services you offer, and exactly how it relates to their problems. When they have a better reality on you and your services, the chance of their becoming upset with your fees is lessened.

There are many situations where a high-priced technical consultant can walk into a situation and give enough pertinent information in one hour to pay for his entire daily fee. If the proper specialist is hired for a program, he can often come into a situation, take a couple of hours to assess it, take a few more hours to ask the proper questions, and then go to work trying to find out where he can contribute the most. Sometimes, after only two or three days, he can come to conclusions which must be summarized in a report. By dictating a long memo of his recommendations, he can save the project large sums of money and deliver results quickly. For quick results, a specialist should demand a high fee.

There are a large number of engineering, data processing, and scientific consultants who are in the specialist category. Often they have a low visibility and their contracts come from a small number of people who know them. Many of these consultants do not even have their business name in the telephone book and yet make an excellent living. The only thing you have to keep in mind is that if you are considering using your specialty and going into high-priced consulting, you're going to have to take the responsibility of educating many of your potential clients as to what you have to offer.

Use a Sliding Scale

Two rates might have to be quoted when dealing with the potential client. If they only want to bring you in for a few days, a high price for the per diem can be stated. For longer periods, you should have a lower, more reasonable fee. There are distinct advantages to taking on longer contracts at lower fees. The most notable of these is that longer contracts will give you time to establish a strong and healthy business rapport with your client and his employees. This is much more important than establishing a quick profit from a high *per diem*. Once a good rapport is established, the client might call you back two or three times a year. With a dozen clients such as these, you could have a complete foundation for your entire business. You should not throw this away by demanding an extremely high figure in the beginning, and thus preventing yourself and your clients from having an excellent future together.

Charge High Fees For "One Time Contracts"

Another situation in which your hourly or per diem rate should be carefully evaluated occurs when you're called in to give advice to a client who will not retain your services in the future. Examples of this are legal situations or insurance claims. If you're brought in to testify in court or to give advice to a certain legal group, it can be the type of advice where you're just brought in on a one-time basis that is very unique to the situation. If you use your regular rates in a situation such as this, there is very little chance of your making money, and it's possible that you will waste valuable time and lose business in other areas. One of your prime concerns in consulting is generating an excellent service which, in turn, produces a satisfied client *with whom you can do business in the future*. If you start doing business with people who will not call you back for new business, there's very little that you've obtained out of your investment of energy and time. Furthermore, very valuable time has been taken away from the operation of your business. A healthy business is concerned with future profits just as much as it is concerned with what's happening in the present circumstances. When you take on work that doesn't generate repeat business, you should demand to be reimbursed to the degree that it is very profitable.

These remarks should be an indication of caution to you. Determining fees is not just a mathematical computation. It includes a mathematical computation of the cost and profits *plus* the consideration of factors such as your determination of what your desire to work with a particular client is and what value you put into a future relationship with him. I would turn down a job which had a momentary but large profit and choose one with only a modest profit if it dealt with a client whom I respected and who was capable of generating a relationship of trust and confidence. Having a mutual trust and confidence in each other is the foundation for forming your future in consulting. If you think it lies in profits only, you're apt to discover the truth the hard way.

TYPES OF COMPENSATION

After a successful sales presentation and proposal activity, you must negotiate the fees for the particular job. You must also determine the method of

TYPES OF COMPENSATION

compensation. The type of compensation can fall under one of the following categories:

1. Per hour or per diem.
2. A retainer.
3. Fixed fees.
4. Cost plus fixed fee (CPFF).
5. Percentage of construction cost.
6. Salary times a multiplier plus fixed fee.
7. Different combinations of the above.

Per Hour or Per Diem Plus Reimbursables

This method of charging for your services is very straightforward. You must first determine what your operating expenses are and what you want for a profit. If you have technical employees, careful consideration of their salaries must also be included.

To establish a per hour or per diem fee, you must first determine what percentage of the time you will be working. Let's say that your estimate is that you will be active in the field 50% of the time. This means that if you're to break even on expenses and you have no other means of income, you must charge at least double the rate you would make if you were working full time. At this point you must also make an estimate of your costs, but mistakes are seldom made in determining what your expenses are; most mistakes are made in estimating the percentage of the time that you will be busy.

Although charging by time seems to be one of the simplest techniques of determining compensation for your services, this is just an apparent simplicity. It's very easy to charge high per diem rates and still lose a considerable amount of money. This can happen if you're not being employed a considerable portion of the time. An inexperienced consultant going into business doesn't realize that a period of profit followed by even a very short period of little work can result in an overall loss rather rapidly.

If you're being retained for a long period of time, you might want to shift to weekly, monthly or annual fees. In these cases be extremely careful of every commitment you make, since once you've made such a commitment, you'll be bound for a long period of time. What looks attractive now might not be so attractive after you've been on the job for a while. Therefore, if someone tries to contract you for a long time on something about which you're unsure, work on a daily or weekly rate for a while until you understand the project. Once you have confidence on what the work entails, and you know what your expenses will be, then feel free to contract out for a longer period of time.

When you're charging on a per diem basis, charge all time that is spent on the project whether it's at the client's facilities, in your office, or anywhere that pertains to the job. Time is even charged for driving to and from the client's facilities. Items for which you can be reimbursed include:

1. Cost of producing deliverables such as charts, graphs, blueprints, etc.
2. Cost incurred by any tests that you make at your facilities or that you pay for directly.
3. Living expenses on jobs away from the office.
4. Traveling expenses.
5. Communication expenses such as telephone, telegraph, postage and package delivery expenses.
6. Equipment rentals or any type of items which were necessary to complete the particular project.
7. Special fees which might have been included in the project, such as legal, accounting or special secretarial services.
8. Computer time.

Charging by a Retainer

When a client wishes to have a consultant's advice or services on hand at all times, a retainer fee schedule is called for. A retainer contract is simply an agreement that the client will pay the consultant a fixed amount of money so that the consultant will be on call and be ready to serve his client on a continual basis. Because the retaining fee is fixed, there is always an upper limit for the amount of time the consultant is obliged to give the client. However, if the client needs more time, the consultant is expected to do his best to provide a service to the client. Therefore, the retainer contract not only assures that the client will have at least a small amount of time with a consultant, but it also gives some assurance to the client that if some extra time is needed, the consultant will give his client a high priority.

From the consultant's point of view, there exists a value to the retainer type of contract that might not be too obvious. Many contracts by retainer are not extremely profitable to a consultant, but they serve a function that is extremely valuable to the consultant regarding future business. By going on retainer, a consultant allows a client to call him at any time, and thus a continual line of communication is established. Since the client is paying for the consultant's time, the client will use the consultant at every possible opportunity, and this keeps the consultant in constant communication with the client. This open line of communication is extremely effective.

Sometimes an entire year will go by where a client does not need his consultant for a large project. However, during that time, the client and the consultant can be in close communication with one another on smaller issues. In this way the consultant is always up to date on new developments in the company. It can be seen from this that working on a retainer is an excellent means of establishing a strong and long-lasting relationship with a client and his employees.

Because retainer fees are not always large, they are sometimes avoided by consultants who don't recognize their value of keeping the lines of communication open between themselves and the client. Any consultant should always mention to his client that he can be retained when their

TYPES OF COMPENSATION

contract comes to an end. He should then give the client a description of what his retaining rates are and how the fees are billed. The client might even be surprised to hear that the consultant works on a retainer basis since retainerships are rare in some technical areas.

In some situations it is advisable for a consultant to refuse the offer of a retaining contract. An example is the situation where the client wants to retain a particular employee of the consulting firm, and the principal consultant knows that the person will not always be available because he is already committed to other contracts. There also exists a situation where a consultant knows he or one of his people might be out of town or even out of state for a considerable period of time, and that it will be difficult to satisfy the needs of the client. If this is the case, the consultant should make his circumstances very clear to the client. If the client still wants to retain the consultant, then a short retainer contract should be written, and it should be stated in the contract that it is understood by the client that the consultant will be occasionally out of town or, in some situations, not capable of giving his full attention to the client.

The use of a retainer is very common in legal work, and it should be used if the consultant plans on consulting to law firms and insurance agencies. The consultant can expect calls from such clients to come in rather sporadically, and therefore his time will not be as easily scheduled as when working with clients who have technical practices.

Another area in which retainers can be used is in dealing with smaller companies that cannot afford to have a full-time specialist aboard, but still require some degree of assurance that they can get advice from a specialist when needed. The same can also apply to any technical project that is extended over a long period of time since it can be difficult to estimate when the consultant will be needed. There are many types of projects which often require sporadic requests for a few hours of consulting time, and in such cases, a retainer is a commonly employed method.

The amount charged by consultants in a retaining contract varies considerably. The reason for this is that a retaining contract, more than any other type, is unique to the particular circumstance, and the fees will have to be determined by the specific needs of the client and the availability of the consultant.

The Fixed Fee Method

In areas of work in which the consultant can accurately predict what he has to do to complete a job, the fixed fee method of payment is excellent. It is excellent from the client's point of view as well, because he knows exactly what he's going to get and he knows precisely how much it's going to cost him. This is helpful from a management point of view where there are many variables that have to be dealt with, because putting people on a fixed fee basis will eliminate one of the unknown variables on a project.

The fixed fee is just what it says. It's a predetermined sum which is paid to the consultant for a well-defined set of deliverables that are to be delivered on predetermined dates. By breaking up the deliverables into a sequence of milestones, the client can determine the value and quality of the work which is being performed. He can also pay fixed amounts as the project moves ahead. In this way the consultant does not have to wait until the end of the project to be paid. Types of work that traditionally fall into the fixed fee category are: studies, investigations, and technical services where it is possible to predict how long the various tasks will take.

In bidding for the fixed fee contract, the consultant must be very careful to include all tasks that have to be done and any materials that he must pay for to complete the job. He would be wise to review his estimates with someone and be very careful that he can do the job for the fee. Identify those areas that might create trouble and make allowances for those particular parts in the contract. If this is done correctly, the fixed fee contract can be profitable to both the client and the consultant. In work where the tasks cannot be clearly identified, the fixed fee method definitely should be avoided.

On large projects where the consultant is asked to give a fixed fee quote, he should not hesitate to require a small contract to pay for his time while he determines the necessary tasks that have to be accomplished. Sometimes a great deal of time has to go into an accurate appraisal of how much the job is going to cost, and a client should pay the consultant to do this work.

If the client does not want to pay the consultant for the time involved in preparing the fixed fee estimate, the consultant should be extremely careful and should not proceed unless he knows that he'll obtain the work if his bid falls within a reasonable range. Doing this kind of estimating can sometimes lead the consultant into trouble, especially where he doesn't know his client very well. If he does know his client well, the client should be willing to pay him to do this portion of the work.

After the consultant has determined all the tasks and completed a cost estimate, he then must review the entire cost estimate and its various parts with the client. If the client tries to bargain heavily to bring the fixed fee down, the consultant should be very cautious. If you predict that your profits are going to be as small as 8% to 12%, it won't take very many mistakes before that profit can be completely wiped out. There's nothing wrong with charging a large profit on a fixed fee basis where there are chances of losing money. Many an engineering company has had to close its doors because they thought they could do a particular job for a fixed fee and then found it was impossible. You must leave considerable leeway in your estimates so that you're safe. Remember, the client benefits greatly on a fixed fee contract because he knows in advance what the project is going to cost him, and this predictability is generally worth a considerable amount of money.

Therefore, don't think you're doing your client any great favors by not taking possible problem areas into account. You can hurt your client as well as yourself if you call something incorrectly. If you have a good rapport with him and serve him well, you're not going to do him much good if you have to fold because of your own mistakes in prediction. The key to good prediction is to first have a complete understanding of what you don't know. Know where your blind spots are and then do a thorough investigation and get the costing data required for a careful estimate. Knowing where you're weak is extremely important, and unless you have that ability, you can get severely hurt in a fixed fee contract.

Despite the potential difficulties that I've mentioned, it should be remembered that if you're a well-organized and efficient consulting company, you can make excellent profits on fixed fee contracts. Fixed fee contracts are contracts that award the able and get rid of the incompetent. Therefore, be aware of your talents and of your weaknesses, and use the fixed fee contract to both your advantage and to the advantage of your client.

If you are going to take on government contracts, you'll find that a large percentage of them come under the fixed fee category. In these cases, you must be completely aware of all the government regulations that go along with the fixed fee estimating and pricing. You should also include in your bid a considerable amount of time which is involved in working with the government officials whose job it is to see to it that you fall within the regulations. It takes a considerable amount of experience to deal with government agencies. Therefore, the novice should approach government contracts with considerable caution.

Cost Plus Fixed Fee (CPFF)

Whenever an engineering or scientific project must be conducted with a large number of unpredictable variables, the costing becomes very difficult. The determination of the feasibility of many projects might employ testing, experimentation or the construction of items which have never been built before. An extreme example of this type of work would be in the space industry. Here, with every new rocket or satellite that is launched, equipment must be designed and built to very accurate specifications in areas where such equipment has never existed before. In cases such as these, the fixed fee technique becomes close to impossible. As a solution for this, a type of contract has been developed which is called cost plus fixed fee.

When working on a CPFF contract, the consultant will be paid for the following costs:

1. Salaries (including vacation, sick leave, etc.) of technical personnel plus all associated costs that go along with those salaries such as payroll taxes, unemployment insurance and social security benefits. The salaries of the principals of the consulting firm are also included to the degree that they perform technical work on the project.
2. Non-salary expenses (direct). This type of expense includes all forms of telephone and communication expenses, living and travel allowances, and all work or costs that are directly related to the project. Examples of the project related costs could be equipment rentals, consultant fees, laboratory costs, various supplies, and reproduction costs.
3. Overhead, which includes: office expenses, library expenses, taxes, promotion expenses and administrative costs. Administrative costs include executive, legal, accounting and clerical salaries, and most salaries which are not directly included on the technical side of the ledger.

A fixed fee is an additional payment that is made to the consulting company, and is obtained by taking a percentage of the cost of the project. Often this percentage is the main issue for negotiation. For small projects this percentage can be as high as 15 or 20%; on large projects it is usually considerably less.

The major problems that can occur in a cost plus fixed fee type project are caused by poor planning. Usually, projects that call for this type of contract have a large number of unknowns that must be dealt with. In case of low predictability, use the CPFF contract to protect your relationship with your client. If the consultant must deal with a client and a problem he does not fully understand, the consulting company can be in for a great deal of trouble. The objective is to make a profit, deliver a good product and also maintain a satisfying and successful relationship with your client. If your client finds that he's spending much more than he was originally led to believe, your good relationship with him can deteriorate rapidly. One of the ways that projects such as this can avoid financial disaster is to have studies done which basically collect the necessary data to increase the level of predictability.

As in fixed fee contracts, the consultant must have a complete company record of all information that pertains to expenditures. This has to be extremely well-documented and very carefully recorded. The accounting techniques used should be extremely accurate, and one must be able to communicate the reasons for the various costs incurred.

This type of contract might not appear to be important to the small consulting company or to the independent consultant, however, it should be recognized that even the independent consultant must be familiar with all the problems that are associated with this type of contract, because he will very often consult to clients who are working on cost plus fixed fee contracts. If this is the case, he must also have very good records of how he spent his time and all the expenses that he generates so that his client can have access to those figures whenever they're required.

Percentage of Construction Cost

In this type of contract, the consultant is paid a particular set percentage of the total construction cost, or a percentage of a well-defined portion of the construction cost. The percentage that is paid will depend on the type of work and the size of the construction job, and of course, these rates decline with the size of the contract. They can start as high as 20% for small construction jobs and drop considerably for large construction projects.

There are some serious problems associated with this type of payment. One of the most obvious disadvantages is that on the average, everyone working on the projects is rewarded equally. With this method of payment, if you have a consulting company which takes pride in high-quality work and if you deliver what you promise, you will be dependent on other less-efficient companies for your profits. There is also no incentive for any of the contractors to try to save the client money. Further, working on this type of payment method can be complicated. These contracts usually depend on rate schedules determined by other groups that might not apply correctly to the particular project that you're bidding for. If this is the case, you could wind up doing a considerable amount of work for what could be a very small amount of return. It is for these reasons that I recommend extreme caution with this type of contract.

However, if the percentage of construction cost method is applied to a well-defined and much smaller part of the entire construction job, the dangers can be lessened considerably. In this case, the consulting company often can have a considerable influence on the outcome of their particular section of the contract and have that one section of the contract succeed. Therefore, in order for this method to be successful, one must define precisely what has to be done on various phases of the work, and strict assignments of jobs have to be adhered to closely.

Salary Times a Multiplying Factor

In this method, payment of the salary of the technical personnel involved in the project is multiplied by the factor which is related to the consulting firm's overhead. As an example, as a member of a consulting firm, I might tell the prospective client that I receive a salary of $200 per day from my consulting firm, and so my firm wants $400 per day for my services. The multiplying factor can vary slightly from industry to industry, but usually falls somewhere between 2 and 3.

Even the definition of what constitutes the salary can vary slightly. However, it usually includes the salary plus all benefits such as social security and unemployment compensation. The extra charges on this salary due to factors such as these should not be ignored, because they can constitute an extra 33% of the basic salary figure.

Small consulting firms that are just starting up are often not quite sure on how to charge for their fees. One excellent way is to determine set salaries for your technical personnel and then choose a multiplying factor that lies within the range of your particular industry. You'll find if you use an average multiplying factor that the final fees which you arrive at will be fairly close to fees that you'd arrive at using other techniques.

Sealed Bid Contracts

Some of you might have been saying while reading this—well, how about sealed bid contracts where only the price counts? My answer to this is very straightforward—if you're in an area where only the sealed bid matters and not the quality of the work, then you're in a troubled industry. You're going to have to get yourself a set of clients where this is not the mainstay of your business. If you're being forced to lower the quality of your work because other people are undercutting you, then you're headed in the direction of many other consulting firms, and that direction is right down the tube.

My advice is to do anything in your power not to be controlled by sealed bid type work and competition that enters bids which are either unethical or poorly thought out. Let somebody else win the contract and deliver a poor product or service. Remember, some of your competition does not know what they're doing, and many of them will not be around a few years from now. Let them have the big job, and let them think they're winning. Build slowly and confidently with an excellent product and satisfied customers who will give you repeat business.

RECOMMENDED TECHNICAL FEE SCHEDULES

In some areas of the technical professions, technical societies publish guides to fees. These schedules recommend time charges for particular types of work or certain fee percentages used in the method of percent of construction payment. Such schedules are recommendations and should not be interpreted as any more than a recommendation. Each job has its own unique aspects, and the fees that are charged for particular work must reflect the specifics of that particular project. However, there are many areas where such schedules can be exceedingly valuable.

Often a consulting company specializes in a rather narrow area and must hire other specialists to do some work that is outside their expertise. The schedules can give an idea as to what kind of fees are charged for those kinds of jobs. Likewise, if an engineer or data processor is interested in consulting, but doesn't know what fees are being charged, he can use these recommended fees as averages to determine what he might expect if he consults in that area.

There are many jobs that will require the use of much higher or much lower fees than are recommended in these particular schedules. If you have a client who asks you why you are not charging a fee that is near what is recommended in the schedule, you should be patient and explain to him how your job differs from the average. There's nothing wrong with a client asking you how your work differs from the average. If you put yourself in the same situation of hiring another specialist to work for you, you too would find it wise to ask him why he's charging you a higher or lower fee than what is in the recommended schedules. These schedules are actually very valuable as long as it is kept in mind that they are merely recommendations.

A problem associated with these recommended fees has arisen over the past few years as inflation and heavy price fluctuations have shown that some fee schedules cannot be used effectively. In particular parts of the country there are shortages of manpower and materials. Rates can vary to a tremendous degree. A good example of this would be construction and consulting engineering fees in Southern California as compared with the state of Maine. In such a situation, one might as well compare two different countries.

As long as one recognizes that he's working with averages (and associated with every average there exists a variance), then there should be no problem in using these schedules. As inflation grows, and costing becomes more of a problem, the averages will still be a good concept, but the variances associated with those averages get much larger and, therefore, the usefulness and applicability of the recommended fees will decrease.

WHEN ARE YOUR OWN FEE SCHEDULES APPROPRIATE?

In some types of consulting such as business consulting, consultants employ standard rates. Usually, these rates are on a per hour or per diem basis, and it's customary for the consultant to publish a rate schedule. This is usually a statement of what their hourly and per diem rates are and how the rate decreases if they're to be employed for a longer period of time. It also includes rates for the different consultants that are in the firm. For example, the rate for a senior consultant will be considerably higher than the rate for one of the junior consultants. In technical consulting, such fee schedules are rarely used. The reason for this is that the fees charged are project-dependent and can vary considerably.

However, if you're a technical consultant who specializes in a particular area and do not work on hardware-related projects, or if you're a scientific consultant, then you might have a fee that is standard and not subject to fluctuation. If this is so, it's beneficial to have a short description of your charges ready to show your potential client. It can also be sent in the mail if someone requires information about your services. On this schedule you should state what your hourly or per diem rates are for various tasks. If your rate drops on longer projects, you should give an indication of how much time will be required before your rates decrease and what your new rates will be. You should have a sentence on the rate sheet which points out that this fee does not apply to all circumstances and that depending on the project, the fee could go up or down. You might also state that the method of payment can also be arranged by negotiation.

Additional Notes On Fee Schedules

Different rates can be charged while you're traveling on longer trips. For example, on some consulting contracts a consultant is paid his regular fee for traveling to and from his client's facilities. Because such travel can sometimes take a considerable amount of time, there might be a drop in the rate in those cases. As an example, if you're planning on traveling across country or to Europe, you should definitely have some kind of arrangement whereby it doesn't cost your client $65 an hour while you're traveling. However, don't get too generous. Your time is valuable, and businesses that treat their clients like charities don't last long.

For local clients, you should indicate on your rate schedule that you do charge for the time traveled to and from the client. However, if the job becomes an extended one, then this hourly traveling charge can be dropped. Standard mileage fees are not dropped under any circumstances.

Sometimes travel time can be exceedingly long, and going to see a client for just a couple of hours can really take an entire day. In such circumstances, you should explain to the client exactly how you're going to charge for this. Going to see a client on a one-day basis only can be very expensive to you, especially if you're not charging for all the time that you travel. Some consultants don't like to charge for time they're not spending with the client. If this is the case, what you should do is charge a much higher fee for a short one-day service. This higher fee will accommodate any losses incurred while you're traveling, and your customer will be satisfied because he knows exactly what he's going to be charged.

You should be certain to explain to your client any other charges that are normally charged to the customer, such as telephone and production costs directly related to the project, reproduction costs for any reports, etc. Other items such as laboratory testing or specialty-type work should also be included in the fee schedule.

Your prospects should know how much time they will need in advance to place you into his schedule. If you are very busy and require booking in advance, then this should definitely be mentioned to the client. Also, if he decides to make an appointment with you and then breaks it, he should be charged a minimum fee. After all, he has interrupted your schedule and this costs you money. The fee that you charge for breaking the appointment should be mentioned on your fee schedule if you plan to enforce this.

Hourly or *per diem* rates should also be included for any of your employees or associates. They might have a considerably different level of technical competence and their rates could differ significantly. If this is the case, you should put a short description of the type of work that is done by them along with what rates are charged. This will assist the potential client and also give your company's expertise a little bit more exposure.

COST BREAKDOWN

In most cases, as a consultant, you do not have to break down your fees for your customer. Simply billing him at a fixed hourly or daily rate, or on a fixed fee basis is usually sufficient. But in some cases your client might wish to know what percentage of your fees are for direct expenses and what percentage is for your overhead. In this section a description is given of some of the cost breakdown terms you should know.

Direct (Payroll)

Direct payroll is composed of any salaries or wages which are paid to staff and officers to satisfy the requirements of a contract. When placing people into various payroll categories, they are categorized using an hourly fee (even if they are on salary). For salaried personnel, one just takes their annual salary (do not include benefits) for the year and divides it by 2080 working hours to get the average working rate for that particular category.

If hours are to be charged that will be above the 40-hour work week, then arrangements (in the contract) have to be made on how the overtime rates are handled. For technical consulting, time-and-a-half rates are customary.

Direct (Non-Payroll) Costs

The direct (non-payroll) cost is any non-salary related cost which a company incurs in order to meet the objectives of the contract. Examples of

COST BREAKDOWN

direct (non-payroll) costs would be:

1. Equipment that must be bought specifically for the job.
2. Equipment rentals.
3. Engineers and technical staff (job shop).
4. Technical consultants.
5. Communication costs directly related to the project.
6. Computer services that are project related.
7. Travel & living expenses related to the program.
8. Special services directly related to the project, such as accounting expenses, program costs, legal fees and laboratory charges.
9. Drafting supplies.
10. Printing and reproduction costs for reports and blueprints and other similar items.

Indirect Costs

The difference between direct and indirect costs is how they relate to a particular project. An indirect cost is incurred for an activity or item that does not relate directly to a project. One method of determining the indirect cost is to add up all the costs that are directly related to the program that the company is working on and subtract that amount from the rest of the costs. However, it is often more beneficial to separate the indirect costs into two parts. Those two parts are the indirect (payroll connected costs) and the indirect (general and administrative—G&A) costs. The sum of these two costs is referred to as the consulting firm's overhead.

The indirect (payroll connected) costs are all those costs to the consulting firm which are directly connected with wages and salaries. A list of a few of these costs is as follows:

Social security taxes.
Federal and State unemployment taxes/insurance.
Workmen's compensation taxes/insurance.
Health insurance.
Life insurance.
Vacation, holiday, and sick pay.
Leave-of-absence pay.
Pension and profit sharing.

Note that these costs can go as high as 30 or 35% and they form a substantial cost to the entire operation.

The indirect (general and administrative) costs are all the costs that are associated with the administrative side of the business. Examples include:

Administrative payroll.
Administrative travel and living expenses.

Business taxes which are not related to income and payroll taxes.
Contributions.
Depreciation.
Advertising.
Insurance which is not related to payroll such as professional liability insurance.
Motor vehicle operation and maintenance.
Office furniture and maintenance.
Computer facility expenses.
Technical equipment and maintenance.
Communication expenses that are non-project related.
Interest charges on loans.
Registration and license fees.
Research and development.
Professional society membership fees.
General office supplies and related office costs.
Accounting costs.
General legal costs.

If you're planning to work with the government, you must obtain information from your government contracting agency on what items you can place in particular categories.

Overhead

Overhead is obtained by adding the indirect (payroll connected) costs to the indirect (G&A) costs. Sometimes this total overhead figure is not a sufficiently thorough breakdown, and you'll be asked what part of your overhead is related to payroll and what part of your overhead is related to G&A. In this case, the items referred to are G&A overhead and payroll connected overhead.

Cost of Services

Your cost of services is the sum of your direct payroll costs, your payroll connected overhead, your G&A overhead, and your direct (non-payroll) costs. This cost of services is not the cost to the customer, but the cost of services to your company. The cost to the customer, or the fee, is this cost of service plus a profit factor which you must negotiate with the client.

9
Negotiation

Prior to any negotiation you should decide whether you want to deal with a particular prospective client. Likewise, prior to negotiation your prospective client should have made the decision that he wants to work with you. If either you or your prospective client have doubts about one another, you will run into trouble during the negotiating phase. However, if you are both sold on each other, you will find that when it comes time to negotiate, all will go smoothly. If you have completed your preliminary promoting and closing steps properly, you are now at a point where you can enter into a negotiation and come to an agreement on fees, methods of payment, equipment, manpower, facilities, etc.

It almost goes without saying that negotiation is a crucial element of running a business. Yet there are certain fundamentals which, if omitted, make for very unpleasant and unproductive negotiations. This chapter shows you how to handle this vital stage properly.

WHAT IS NEGOTIATION?

Negotiation is not necessarily an adversary situation, but it can become so when the preliminary steps to negotiation have not been handled properly. In any successful negotiation between a client and a consultant, both parties should be satisfied with the conclusion reached; a successful negotiation is not one party winning and the other losing. The goal of a successful negotiation is to form agreements which will assist both parties to reach their objectives. The purpose of negotiating is to provide a rational framework where business agreements can be made, so that the objectives of one party do not clash with the objectives of the other. If upsets or clashes occur after you and your client are working together, it is probable that the negotiations weren't handled properly.

It's good to remember that the ideal negotiation does not occur between two groups. Ideally, a negotiation occurs between two individuals. Therefore, if you are a small company, you should have someone in your group who is a negotiator and who, at critical moments, has the final word. If you negotiate properly with a client, you will maximize the efficiency and harmony of your relationship with one another.

There exists a negative type of negotiation in which one of the parties is not concerned with creating agreements that are mutually beneficial to everyone. In this type of negotiation one of the parties is looking out only for himself. This is the type of situation that could become an adversary type of negotiation. Ideally you should be able to avoid this type of arrangement by choosing your prospective clients properly. Unfortunately, you sometimes find out about your prospective client's characteristics only during the negotiating phase. If this is the situation, don't despair, because it is better that you find out about your prospective client's negative characteristics before you sign a contract than after the agreement is made.

A common pitfall in negotiation occurs where one side does not understand the objectives of the other side. It is essential that you have a complete understanding of your prospective client's objectives and that he is fully aware of your objectives. If you have chosen your prospective clients properly, you will find that the negotiating phase is done so quickly and easily that you might not even notice it has happened. In most technical consulting the jobs are very straightforward and the rates and adjustments to get the job done are easily agreed to. Even though most negotiations go very smoothly, it is still necessary for you to understand some of the fundamentals about making agreements so you will avoid potential problems.

NEGOTIATING YOUR FEE

While negotiating, one should have a guideline as to how much various items will cost. These should have been prepared previously during the sales presentation and the proposal activities. For smaller jobs, this is not the case, and you'll have to start working out the various costs after you've been told by the client that he wants you to do the work. In determining the method of compensation, many factors must come into account. Most important are the objectives of the client and his type of program or project. Often, the type of project dictates the type of payment method.

Other factors that can determine the type of payment are the accounting procedures used by both the client and the consulting company, as well as any particular constraints that are placed on the project. An example of such a constraint would be that all or part of the work is under the auspices of the government and that many of their rules will dictate how particular compensation will be carried out.

A Critical Point

The consultant should never forget that the negotiation is one of the critical points in the entire program. He must work with the method of compensation agreed to, and he must obtain a fee that yields a reasonable profit. It's during this negotiation point where he'll learn a lot about his client. He'll be able to see how his client reacts to the figures which he presents as approximations to do the job. Negotiations with an inexperienced client can be difficult. However, if you use patience and explain the reason for all your figures, you can avoid a large number of problems.

During the negotiation the client can try to bring the consultant's fee down. Although it is praiseworthy that he's trying to save his company some money, the consultant should be careful not to agree with the client. Don't lower the quality of service down to accomodate a client who wants lower fees. When the job is over, people will forget exactly how much was paid, but no one will forget the quality of your work. Success of your business will depend on your reputation which, of course, is strongly dependent on the quality of work. This does not mean that you have to be a perfectionist, but it does mean that you should deliver an excellent product. If you start decreasing the quality of your product just to save a few dollars in negotiation, your consulting work and reputation can be severely damaged.

Be Firm

Be firm during the negotiations. If you have to, let someone else do a low-caliber job. Do good work and be proud of it. It's this excellent work which will be your main method of obtaining more work in the future. There's a great reward in sticking to your guns and only doing high-quality work. After you've been a high-quality firm for awhile, you'll find prospects approaching you because of your reputation. They know you'll do a good job. They also know that you might charge a little bit more, but they know that when they deal with you they're going to be getting the results they want. When this happens, negotiations become much easier, and you're on the road to working with and for the right kind of people.

High Quality Work Is Not Inexpensive

To do a high-quality job requires the extra energy and attention that does cost money. However, it's often a misunderstood area. You just don't do a high-quality job by dumping more money into a project. You really get high-quality work by putting a little bit of extra money into having a well-organized company and extremely well-trained and competent people. Then, you don't have to force high quality. The extra money invested into your organization will show high-quality results coming out naturally, instead of by force.

Don't Be a Pawn

Another thing to remember is that some clients really enjoy the negotiation part of their new work. They like sitting down and playing the game of seeing who can save the most for his company. Although that's not my idea of negotiation, there are some people who feel this way. Don't be flustered just because you have a client putting a little bit of pressure on you to bring your prices down. Stick to what you believe is right, and give in only on areas in which you know you can afford to do so. State your fee for doing an excellent job and then do an excellent job. Otherwise, don't take on the work.

He's Making a Profit

I want to stress a point which should never be forgotten while you're sitting at the negotiating table with a client. If you deliver a good product and he knows it, he's getting an excellent deal from you. Don't forget for one moment that he's going to make a profit from the work that you do. Just make sure you understand this while you're negotiating. *He's going to make a profit on what you do.* If he doesn't, he's not a businessman and he's in trouble. Generally speaking, the work you do will somehow be charged to somebody else with a profit on top of it. So, what's really happening when you negotiate is that he's negotiating for how large a profit he's going to make off of your work. Not how much it's going to cost him, but how large a profit he'll receive when he, in effect, sells the results of your work to someone else. What he's really negotiating for is not how much your work costs, but how much it costs him relative to how much it's going to be sold for. If you don't lose sight of this, it will become much easier to stick to what you believe and not be swayed by any false accusations about how much your service costs.

He's Saving Money

By hiring you, he doesn't have to pay the expense of having experts in his organization continually. He also saves money on any down time that would be incurred by not being able to keep a specialized employee busy on a continual basis. Remember, the client who knows his business often needs a good consultant. In fact, he may need many of them. So, you're providing an excellent service, and you should definitely receive a fair return for your own investment and the excellent service that you produce.

Make a Healthy Profit

Businesses that succeed make healthy profits. There's nothing wrong with making a good living in exchange for delivering an excellent service. That's part of the reason for being in business. A good client will respect you for making the money you deserve, and he will want you and your business to prosper. Charging three times the hourly salary rate is not high and will not be considered so by your client. Let him know that he's paying for quality service and charge accordingly.

Making Estimates

Prior to or during the negotiations you may be asked to make estimates on how much something will cost. A good habit to get into is: *Never give an estimate as a number alone*. For example, if a client asks how much a particular item will cost in the construction, don't come back with an amount such as $10,500. On an estimate number, you should come back with something such as $10,500 ± $700. If he doesn't like this, he can ask you to refine your estimate, at which time you can go back and return with a quote such as $10,500 ± $400. Therefore, all estimates should be given in two numbers: the central price about which you think the cost will fall, and the spread (or uncertainty) about that price. This is an excellent means of communicating to the customer what is involved in the estimate. It's also amazing how few technical professionals use this means of communicating estimates. When somebody asks you for an estimate and you only give them one particular number, this gives them very little information.

When somebody asks you for an approximation, you should give the approximation to the client on a sheet of paper. It doesn't have to be formal or typewritten. It can be handwritten on a sheet of paper and be included in the negotiation notes. Also, if you don't want to get as formal as handing him a piece of paper, you should always have a notebook present when you're discussing fees—then you can give him the verbal quote and jot down the quote in your notebook. This avoids possible problems that could come up a few days later. It's very easy to give an approximation and later on have somebody use that number as the accurate figure or the figure that they plan to go by. Don't let this happen. If somebody does say, "Well, I thought you said it was so much," you can always take out your notebook and state that what you gave them was an *approximation*. Keeping records such as this can save not only time, but also prevent a few unnecessary disagreements.

The Specialist

If you're a specialist who charges fees that may be considered high by some clients, you may be asked why they're so high. If you're a specialist and you have very little competition, an excellent answer to why your fees are where they are would be to politely ask, "High compared to what?" You'll be very surprised when the client recognizes that there's no one else who can really do the job the way you can and, at that point, his question should be answered.

The Beginner

If you're new to consulting, you may have the idea that you should charge a substantially lower price for your consulting just to get a job to get you going. This can be a great mistake. In determining your fee you should find out what the average is in the industry and charge a fee near or above that average. If you start by charging a fee that's too low, you might only get clients who will only take on extremely low-priced help. Clients such as this can be a disaster in consulting, and there's no reason to work for them. Charge a little above average and find out how you do. The chances are that you have a lot more experience than you believe you do and you'll do an excellent job. Remember, you're going into business, and you have to charge a fee that's going to result in a profit. One thing's for sure, you're guaranteed not to expand if you start by charging too low a consulting fee.

One other thing that you need to know is that when you charge what you consider to be a high fee, the majority of the businessmen you deal with won't consider that it's high at all. In fact, you could work many years in consulting without anyone even questioning your fee. Forty-five dollars per hour is an average fee today, and many businessmen are used to paying much higher fees for consultants. Remember, a good businessman doesn't look at just your fee. He looks at what he's paying for and what he's getting, and then judges the difference between the two.

In some situations a client who does not know you will ask you how much a job costs before you even get into the details of what the job is about. I suggest that you don't give him any details with regard to cost until you see what the job is. However, if he insists, you can give him a rough approximation of what you've received on similar jobs. You should continue stressing the fact that on the surface they appear to be similar jobs and that, if indeed his project is like the others, your fee will probably fall in the range quoted. However, the proper time to discuss fees is after the problems are understood and the client has decided that he's interested in having you deliver your service with the type of quality that he believes you're capable of producing.

Don't Rush It

When you are asked for certain figures, you are under no obligation to tell a client how much something will cost immediately. You can politely tell the client that you can have those figures ready in a couple of days. You can then go back to your office to do your homework and find out just how much the job will cost you. You must do a very careful study on all possible cases involved. In fact, if a client watches a consultant give quotes right off the top of his head without any thought, the client, if he's a wise businessman, will think less of him, knowing that the consultant is not thinking in advance with enough preparation. So, it's beneficial to you to go back and do your homework, summarize what your costs and profits are, and then go back to your client. Even if it's a small job, he'll have more respect for you when he sees that you've thought the problem out in sufficient detail.

Anticipating Additional Charges

From the first moment the consultant is brought in to assess the client's needs, he should constantly be making a list of items that could cause a possible problem, delay or extra expense to the program. If he keeps this list in a notebook, he might be surprised to see how large it starts to get. This list is going to be extremely important when he adds all his known costs and all his *unknown* costs. These unpredictable areas have to be compensated for in any quotation to the client and, unless you have a list, you're not going to be able to include the proper variables. It only takes

OFFERS, ACCEPTANCES, AND COUNTER OFFERS

missing a few important items to have your profit disappear. Therefore, be careful and have this list ready.

You can use this list for a second and a much more important point as well. After you've given your quote to the client and he has come to an agreement with you, you must then form a consulting agreement or contract. Any areas where problems, delays, or losses could occur should be mentioned in the contract. If there are any additional costs for handling these particular areas, the client must agree to pick them up. A good way of handling this is to mention the possible problem areas and state in the contract that minor revisions to the contract are anticipated on these points. Then, when you later ask the client to revise the contract in those particular areas, he will already have an understanding that some of these problems were a possibility. The chances are he will not be surprised and will be very cooperative instead of being upset. Chapter Ten has a more in-depth treatment of this subject.

By being this thorough, you'll also be doing your client a great service. It's by proper anticipation of problems that he'll be able to make his project succeed. He will also get a different viewpoint of the value of your services. He'll begin to see that your candid approach is saving him time and money and is allowing him to run his own business more efficiently.

OFFERS, ACCEPTANCES, AND COUNTER OFFERS

The Offer

An offer is an indication on the part of one party of his willingness to contract with another party on terms which are stated by the offeror. The offeror states his willingness to give something in return for a promised act on the part of the offeree.

Let's say that you stated to a potential client that you will deliver a set of engineering reports to him if he promises to pay you ten thousand dollars in return. In this case, you have made an offer. You have stated your willingness to exchange a certain set of goods in return for his promise to pay you the ten thousand dollars. Note that in this example you didn't state that you were hopeful that you could deliver the engineering reports. You stated that you would. This is an important part of an offer. An offer is a statement of your willingness to follow through with your part of the exchange. Having the desire to fulfill your part of the exchange, or just being hopeful that you can do it, does not constitute a valid offer. This is an important point in engineering consultation. For example, if you're dealing with a prospective client and you've been speaking on a preliminary negotiation level, you must realize that preliminary negotiations do not constitute an offer on his part. If he's just finding out what your capabilities are and how you might fit in on his particular project, he has not necessarily made an offer. If he does make an offer, you should make sure that all the details have been completely worked out. In most circumstances in technical consulting, contracts are written in which many of the details have not been worked out. In these situations, it is helpful to state in the contract that certain points have yet to be finalized so there is no question at a later date.

The offer may be made in many ways. The offer may come in the form of a proposal, which could be either unsolicited or at the request of a potential client. It could also come in the form of a bid which is one of many bids in response to a potential client's request for a quote.

The Acceptance of the Offer

Once a valid offer has been made, the offeree now has the power to form a legally enforceable contract by communicating a valid acceptance. A valid acceptance cannot be taken back by the offeree. The acceptance is legally binding. For acceptance to be valid, it should correspond with the conditions of the original offer. If it does not meet the conditions of the original offer, or adds conditions to the original offer, then the acceptance is not valid, though it may be deemed a counter offer by the original offeror.

In the technical professions, where the initial offer can be sometimes misunderstood for technical reasons, the offeree must ask for more information with regard to the content or terms of the original offer. Any such inquiries are not counter offers or acceptances. Therefore, if you make a proposal to a potential client and he comes back and inquires about price or content, you should make sure that you don't interpret his remarks and questions as possible counter offers but merely as inquiries.

If the original offer was made to a particular person, company or agency, then that particular party must be the party which accepts the offer. However, there are general offers to the public which may be accepted by anyone who can fulfill the conditions of the offer. For example, if an engineering company made a public offer stating that they would pay anyone $1,000 for information which solved a stated problem, then anyone who provided the proper information could accept the offer. In this situation any number of people who provided the proper information would yield valid acceptances.

Of course, if the potential client initially rejects your offer, then this constitutes a termination of the offer, and he cannot change his mind later and send you a legally binding acceptance. Also note that for any rejection to be valid, that rejection must be indicated to the offeror and not to anyone else.

Note that an offer only remains valid provided that the conditions stated in the offer continue to exist. For example, let's say that you offer to do a particular job for a prospective client, provided that you can retain the services of another consultant besides yourself. You make such an offer, and then three or four weeks later you find out that the other consultant will not be available to you. At this point the new condition is communicated to the offeree, and because the conditions have changed, the original offer is no longer valid.

Generally speaking, an offer can be revoked by the offeror during the time the offeree is deciding whether or not he should make an acceptance. Therefore, if you make an offer to a potential client and then determine in the meantime for one reason or another that you do not want to carry out your offer, you are legally entitled to call him and revoke your original offer.

During negotiations, an approximate price quotation does not constitute an offer. Therefore, when you are negotiating with a potential client and he asks you how much various items will cost, you must be careful of how you give your answer. If it's an approximation and not an offer, you should say so. Of course, a quotation can become a valid offer when the quoter's intention is such that he is willing to deliver a particular set of goods or services for a particular price.

If an offer is given with a termination time and that time is reached, then the offer terminates. If no time was stated, the offer will terminate after a reasonable amount of time. In consulting you might find yourself making offers that could get you into scheduling problems if you don't stipulate the termination time along with your offer. Don't assume that the potential client to whom you're making an offer will give you an answer in a time that seems reasonable to you. You might expect an answer within a day or two, and he could call you up a week later saying that he accepts your offer.

Counter Offers

A counter offer is simply a proposal or a new offer made by the offeree. In this circumstance, the offeree turns into the offeror, and the original offeror is now the offeree. For example, if you promised to deliver an engineering service for $10,000 and a potential client responds by saying that he would make the agreement for $8,500, you now have a circumstance of a counter offer. Your potential client is now the offeror and you are the offeree. Note that the counter offer is not an acceptance of your original offer, but is regarded as another offer which you may or may not accept. When you are working in a negotiation stage with a potential client and you give him a price quotation which is just an approximation and not an offer, and he answers back with a counter offer, the proper action is to inform him that your original statement was not an offer but merely a price approximation. Don't get into a counter offer situation prematurely.

As another example, let's assume that your potential client originally accepted your conditions of doing some work for $10,000 but, in addition, he stated that he would also like you to do some extra work for that fee. Note that this is not an acceptance on his part, but it is again a counter offer.

A Word of Caution

The consultant would be wise to avoid any contractual negotiations or contractual work that is associated with the presence of liquor. A common practice today among businessmen is to talk about details of contracts while having their highballs and martinis. This is poor business practice, and just because it is common in our society doesn't make it a good idea. One of your primary concerns should also be the maintenance of good client relations. Even when people have only one or two drinks, they're partially intoxicated, and that's a very poor way to start a healthy business relationship. If you are placed into a position where your client demands to continue to talk business while having his martini for lunch, simply delay the conversation until a more appropriate time.

PAROLE EVIDENCE

Generally speaking, contracts are valid whether they are oral or written. They can also be valid as the consequence of conduct and action by the parties involved. *The advantage of the written contract over oral contracts cannot be overemphasized.* Often there are disputes as to the original terms of an oral contract. When the contract is written the terms are straightforward, and the contract is the document which remains the stable reference point for both parties. Complications arise if oral agreements have to be established by testimony, since testimony is often conflicting. People's memory cannot be relied upon, and it is customary for the courts to put little emphasis on oral evidence. Courts rely on the written document. This rule of relying on the written document is referred to as **Parole Evidence.**

The Parole Evidence Rule is a rule which refers to words spoken before or at the time a contract is established. The rule is that words spoken before the contract was made will not be allowed to modify or contradict ambiguously written contracts. This assumes that the contract is a complete and enforceable agreement. This rule is important because it affects the way the court will look at the pre-contract negotiation period. Thus, a consultant should take heed of this rule. In consulting, a great deal is discussed in the negotiation period, and if the contract does not represent what was stated in that negotiation period, trouble can arise.

The implications of the Parole Evidence Rule are that if you are negotiating a technical contract, the final written document should accurately portray the conclusions that were arrived at in the negotiations. If any notes or statements were drawn up in the negotiation period, then these notes should be referred to or be made part of the contract. You must include these extra documents with the contract since the Parole Evidence Rule will stop any changes in the contract that are based on oral testimony or other writings that were made before, or at the time, the original contract was established. The Parole Evidence Rule makes an exception when the original contract refers to specific writings or exhibits that were used to help establish a contract. In effect, the Parole Evidence Rule is the basis of regarding these extra documents as part of the original contact.

There are a number of situations in which the Parole Evidence Rule does not apply:

1. The Parole Evidence Rule applies to written contracts that are assumed to be complete. If the contract is incomplete in its form, the Parole Evidence Rule may not apply. Note that it is sometimes difficult to determine if a contract is a complete one. In principle, it should include all the items that both parties originally agreed upon. There are no absolute standards that the court uses to determine if a contract is complete.

2. If there are ambiguities in the contract, then the court may allow the presence of extra evidence. The extra evidence that is entered, however, is usually employed to clear up ambiguities, and not to add or subtract major sections of a contract.

3. If one party used illegal means to induce the other party to enter into the contract, then the Parole Evidence Rule will not prevent the induced party from demonstrating evidence against the other party.

4. If the contract in dispute has sections that are missing, then oral testimony will be allowed to determine information with regard to the omitted sections.

The Parole Evidence Rule is a very practical one since it helps establish a certainty in the final written document. Since the final document is usually the result of a negotiation period in which there might have been many offers and counter offers, we see that the Parole Evidence Rule forces the final document to clearly state what the final agreements were. When you are undergoing your negotiations, you should therefore be constantly making a list of what items should and should not go into the final written contract. Then, when it comes time to collect the data, you'll have the complete list of all the items that you feel are necessary to be included in the contract. This will avoid any possible problems at a later date.

10
Contract Basics for the Consultant

If you wish to operate your own consulting practice, there are some basic principles about contracts with which you must become familiar. In most circumstances a client will ask you to perform a straightforward job, you will agree on the fee, and sign standard contract forms that are used by his firm. Because the process of finding work and making agreements is so straightforward, it is sometimes easy to underestimate the importance of understanding the implications of the agreements that are made. What are your obligations once you have signed on the dotted line?

When I first thought about writing this chapter, my initial inclination was to refer you to a few textbooks on contract law. But unfortunately, these are not written for technical consultants and in general they are overly specialized and confusing with their legal jargon. Therefore, I decided to include a chapter that emphasizes those aspects of contracts that will help you in your own practice. I am not an attorney, and I do not intend to present you with a definitive treatise on contracts. My intention is to introduce you to the language and basics of contract law so you will be able to handle some contract situations yourself, and to converse intelligently with your attorney if the need arises.

Although it is not always necessary to consult an attorney before entering into a contractual agreement, you should not rely on nonprofessional opinion should the need arise for legal advice. This chapter, though a good general introduction to consulting contracts, is not an authoritative legal source. Also remember that laws vary from state to state, and that the interpretation of laws varies from court to court. When in doubt, see an attorney.

The emphasis in this chapter will be on the formal aspects of contracts. However, the reader should keep in mind that the most important aspect of a contract has to do with both parties' intent to meet the duties and obligations that are outlined in the contract. Without this good faith in each other, the contract becomes worthless. Your evaluation of the person with whom you are dealing is more important than the contract. Don't think that the contract will always pull you out of the mud if you get in trouble that arises from dealing with the wrong person. The legal hassles that can arise from dealing with the wrong clients can be large enough to cause you large losses, if only from the fact that your time and attention is taken away from more important matters. A large number of sample contracts is given in the Appendices.

IMPROVING THE CONSULTANT-CLIENT RELATIONSHIP

If you're called in to do even the smallest job, don't hesitate to write up a short statement of work and formalize it with the essentials of a contract structure. Asking your client for a contract is an excellent means of increasing your common understanding with him. Good businessmen are usually anxious to enter into a contractual agreement with a consultant. In fact, if you don't know your client, and he hesitates to enter into a contractual agreement with you, it's a sign that there's something wrong. You must get more information in such a situation and find out exactly where he stands. Further, a written document can be used to avoid misunderstandings as to what each party meant in the negotiation period.

Because the contract often forms the foundation of agreement between the consultant and his client, this document can assist the consultant in keeping his customer satisfied. Once one knows exactly what one wants to do and gets it written down on paper, the communication between the client and the consultant is bound to improve as a result. If you don't operate with a contract, the chances are you won't get into legal difficulties. However, you could wind up with an unhappy client who is not giving you repeat business. There are far too many examples of consultant-client relationships that have gone astray and which could have been saved through the use of a properly written agreement.

RETAIN AN ATTORNEY NOW

It is better to retain and get to know an attorney before you need him. Developing a good rapport with your attorney early will enable you to save time and avoid any losses that could be incurred by dealing with an attorney who is not well suited for your type of work. Do not wait for a problem to arise before you start a relationship with an attorney. Call one and tell him that you are starting a practice and you would like to meet with him to see if he could represent you. Explain that you are seeking an attorney who can handle your type of business transactions and that you don't expect to be charged for your first visit.

It is not difficult to work with attorneys if one is familiar with the few legal basics and with the vocabulary of contractual law that is presented in this chapter. Something as simple as not knowing the legal language is enough to put a barrier between a consultant and his attorney. Without some knowledge and background in the type of work that the attorney must accomplish, the consultant is not able to differentiate between competent and incompetent work until, of course, it's too late.

DEFINITIVE ASPECTS OF THE CONTRACT

The American Law Institute, in its RESTATEMENT OF CONTRACTS (1932) defines a contract as:

> "A promise or a set of promises for the break of which the law gives a remedy, or the performance of which the law in some way recognizes as a duty."

Note that a contract is not just an agreement. It is a specific type of agreement. A contract is an agreement that explicitly or implicitly applies to some form of an obligation. It is usually an exchange of promises and/or assents by two or more persons to do or refrain from a particular act. The agreement is customarily recognized as a contract when the obligation is recognized by authority and enforced through the law.

What Constitutes A Valid Contract?

There are six conditions that a contract must fulfill in order to be enforceable. There must be (1) an agreement, (2) between parties which are deemed competent, (3) founded upon the assent of the parties, (4) supported by a consideration, (5) made within the law, and (6) made for a lawful object.

The Parties to a Contract

There are a few definitions that arise in contract language with regard to the parties involved. These are as follows: The **promissor** is the person who makes the promise, and the **promisee** is the person to whom the promise is made. Once the promise is agreed to, then the promissor is obligated to satisfy his side of the agreement, and he is sometimes called the **obligor.** In such a promissory arrangement, the person who receives the benefit from the promissor is referred to as the **obligee.** Note that the parties to a contract do not have to be individuals, but could be a partnership or even an entire corporation. A government or government agency can also be one of the parties to the contract.

Valid Agreement

For a contract to be binding there must be a valid offer to a party which makes a valid acceptance of that offer. If either the valid offer or acceptance is absent, then there is no agreement and there is no correspondingly enforceable contract. Another condition that is necessary to the validity of a contract is what is referred to as a **meeting of the minds.** This could be defined as a direct and clear communication between the parties. Simply speaking, both parties must understand each others' intentions.

Competent Parties

In order for a contract to be enforceable, there must be a competent promissor and a competent promisee, each of whom is legally capable of meeting his obligation in any contractual agreement. It should be clear that all persons do not have equal legal capacity to enter into contractual agreements. A person under a minimum age will be required by law to refrain from entering into a contractual relationship until he reaches the acceptable age.

Under no circumstances can a person contract with himself. Even if he holds several posts of responsibility, he's still not in a position to maintain a contract with himself. A contract is always with two or more distinct parties.

If you enter into a contract with a person who is intoxicated, the contract remains binding provided that he knew that he was making a contract, even though he would have not have made such an agreement had he been in a sober state (or unless it can be proved that the other party to the contract induced him into a state of intoxication). If the person was intoxicated to the degree that he did not know he was executing a contract, then there is no validity to the contract, and the court looks upon the person who was intoxicated as if he were insane at the time he went into agreement.

Reality and Genuineness of Assent

In order for an agreement to become an enforceable contract, it must be considered genuine or real. Just because two parties have entered into a contract does not mean that the agreement is real. Ideally, both parties understand the subject matter and the law completely, and they are agreeing to the same thing. However, in a situation where one or both parties have different or mistaken understandings about the subject matter or a particular item within the contract, then any agreement they might have had could be unenforceable.

If there are mistakes of fact that are made by one or both parties, the contract can in some circumstances be invalidated. (There are different types of "mistakes" covered in the next chapter, and the reader should know what they are and which particular responsibilities he has when he has signed the contract.)

Valid Consideration

A **consideration** is something that is deemed valuable and is exchanged between the parties named in the contract. Each party to the contract must offer a legally adequate consideration to the other party in order to create a valid contract. If one of the parties does not give consideration, then the law would claim that one of the parties was merely intending to be the recipients of a gift, and in such a circumstance, a contract would not be considered to be enforceable. Thus, to have a valid consideration, the exchange which is stated or implied in the contract must be present and occur in any of the following forms:

1. A promise for a promise.

DEFINITIVE ASPECTS OF THE CONTRACT

2. An act for a promise.
3. A forbearance for a promise.
4. A modification for a promise.
5. The modification or destruction of a legal agreement in exchange for the creation of a new legal agreement.

In consulting, examples of exchanging a promise for a promise are by far the most common. For example, if you as a consultant promise to deliver a certain set of engineering documents to a client in return for his promise to pay you $1,000, then you have formed mutual promises that are a sufficient consideration in the eyes of the law.

In most consulting situations the consideration is the performance of an act. However, a legitimate consideration could also be the promise not to act or not to deliver, i.e., a forbearance. If you wished to buy a promise from another party that he does not act, you would be exchanging your promise for his forbearance. Examples of this would be somebody's giving up their right to property or their right to sue for damages. A typical forbearance occurs when someone gives up the right to collect money for the promise that the funds will be delivered at a later date. A contract constituted by a forbearance for a promise can also occur in the consulting business. If someone owes your business $2,000, and you refrain from collecting the debt in exchange for a promise from him that he pay you the $2,000 in one month, you have created a forbearance in exchange for a promise that is also a sufficient consideration to form a contract.

If your client tells you that he'll pay you $1,000 for a set of engineering drawings and you answer by sending him the drawings, you've performed a valid consideration in the form of an act for a promise.

Now let's say that you have a contract with a client to deliver some engineering drawings at the rate of one drawing a day and that during this contract you create a new set of terms—say you mutually decide to deliver one drawing every two days. There exists an exchange of destroying the old contract and incorporating the old and revised terms to form a new contract. This effect of exchanging a new contract for an old one can act also as a valid consideration.

It can be seen from these examples that a consideration is an entity that is gained or lost or exchanged in the course of fulfilling a contract. It does not have to be a material entity in order to be a valid consideration. A consideration is simply something of value that is received by or given at the request of the promissor in exchange for his promise.

In order for a consideration to be valid, it must be a consideration that is understood and accepted as valuable by both parties. An understanding of the consideration must be communicated to both the promissor and the promisee, and they must look upon the consideration as sufficient inducement to carry out their duties and obligations of the contract.

Pre-Existing Duty as Consideration

Let's assume that you are dealing with a client with whom you have a contract to design an electronic circuit. If you now wish to draw up a new contract with him, you cannot offer as consideration in the new contract any acts or duties that are already mentioned in the original contract. If you already are legally obligated under the first contract to deliver an electronic design, any extra bargaining must concern different consideration than that used in the initial agreement. Of course, if the original contract were unenforceable for some reason, then you could form a new agreement with your client on the same consideration.

Some states allow exceptions to these rules, which are very important with regard to technical contracts. If a consultant determines that he cannot complete the original job at the agreed-upon price, he may then wish to go into a new bargaining situation. Some states rule that if he goes into a new bargaining situation and offers the same duties at higher prices, the original contract will *not* be made invalid. If the courts didn't allow situations such as this, consulting contracts would constantly be broken due to needs for renegotiation, since on technical contracts it is sometimes very difficult to estimate the time and money that it would take to do a particular job. Often a company will bid to build a structure based on today's cost, and by the time the structure is started, the price of the goods and materials could have gone up 10 or 15%. In some extreme cases prices have been seen to go up 30 or 40% by the time the contract is initiated. This exception shows us that the consultant should not think that he can go into a contract and change the contract anytime he wishes. If problems such as this are anticipated, he should make sure that there are renegotiation clauses in the contract which will make it possible for him to renegotiate without difficulty.

A matter concerning consideration can arise in technical contracting and consulting when a contractor refuses to complete his construction obligations unless the client promises him some form of extra payment. If the client makes the promise to make the extra payment, the question arises whether the owner's new promise forms a binding agreement. This is very important because if you're doing a particular job and you ask for extra money and you do not sign a new contract, you'll want to know if your agreement with the owner to finish the job forms an enforceable contract. Some courts have ruled that the owner's new promise forms a binding contract. The reasoning that is employed is that the original contract was rescinded by mutual agreement, and that a second contract was executed when the owner promised to pay the contractor the extra sum. Most courts, however, have ruled contrary to this. Most courts recognize that the second promise, made by the owner, was made without consideration and that the old contract is still valid. Thus, if you find yourself in this circumstance, be aware of the fact that your client might not be legally obligated to pay you the extra sum that he has agreed to pay you.

Adequacy of Consideration

Generally speaking, the courts are not concerned with the amount or value of the consideration given or received by the parties. That is, the courts are not concerned with the adequacy of consideration. If you want to enter into a contract of selling your $50,000 computer for $1,000, the court would not declare such a contract void. The law is not concerned with the fairness of a bargain; it is primarily concerned with the fact that a bargain exists. (This, of course, assumes that the agreement is being made to the satisfaction of both parties. If there were circumstances of fraud or duress, the law would take this into account, but these would not be issues of adequacy of consideration.)

On the other hand, there are some situations in which the court is concerned with adequacy. These occur when one party agrees to give the other party an item which is insufficient for a proper exchange. For example, suppose you have a client who owes you $10,000 that was illegally incurred through, let's say, a gambling debt. You then entered into an agreement with him such that you will agree to give up your right to the $10,000 in exchange for his computer. The court would find that this was an inadequate consideration. It was inadequate because no benefit would be received by you because you were giving up the right to collect a debt that was not legally collectable in the first place.

DIFFERENT TYPES OF CONTRACTS

Unilateral and Bilateral Contracts

A fundamental of contracts is that a party or parties make promises which originate from the offeror and are consummated with an acceptance. In a unilateral contract the situation is simplified, and only one party makes a promise. Suppose you're a programmer and offer to sell a set of programs to a potential client for the price of $10,000. You then obtain an acceptance to your offer by receiving a check in the mail for the $10,000. This is a unilateral contract. The person with whom you dealt made no promise at all. The only promise that was made was that you would deliver the drawings for the amount indicated.

Whereas in a unilateral contract one of the parties makes a promise, in the bilateral contract there exists a set of promises. Each party in the contract takes the position of both a promissor and a promisee.

Let's examine a modification of the last example. As before, let's say that you offer to sell your programs for $10,000. In return, the client answers this promise with a promise of his own, which is that if you deliver the programs, he'll pay you the $10,000. In this case, both parties made a promise, and there was in effect an exchange of a promise for another promise. Note how this differs from the unilateral contract example in which you made a promise but you did not get a promise in return; you received money in return instead of a promise of payment.

Express and Implied Contracts

An express contract is one in which the parties have explicitly declared their intentions in terms of the transaction. Their statement can be done orally or in writing. In consulting, an express contract is one that is regularly encountered where there is little doubt about the explicitness of the transaction.

There are some circumstances in which parties do not orally or in writing arrive at an agreement, but through their actions a legal contract was indeed made. **Implied contracts** often occur when an offer has been made and the acceptance was either absent or improper. Suppose, for example, that you were supplying a client with engineering documentation on a regular basis once a week, and that upon delivery you were paid for your services. Then, after many weeks of such activity, you brought your client information that he said he didn't order. In such a situation you could bring him to court, and it's possible that the court would decide in your favor, on the basis that there was an implied acceptance because of your activities with your client in the past.

To take another example, suppose that you delivered certain engineering information to a client and then he used those documents to his benefit. He then informed you that he wasn't going to pay you for the documents that you delivered. In such a situation, the courts often find an implied acceptance, and you would have the legal right to receive your money for the goods that you delivered. In circumstances where your client did not receive any benefits from the documents that you delivered, the case would certainly be less clear, and the specific details of the situation probably would determine how the court would decide.

The major difference between implied contracts and express contracts is that in the implied contract, the oral or written declaration of the intentions of the parties is absent. Instead of having an oral or written declaration of intention, the intention becomes evident by the conduct and action of the parties. Let's say that in the normal course of a business day you indicate to someone that you have some equipment that would be of benefit to his project. He indicates that he wants the equipment, and you send it to his shop with a bill. If he uses these supplies to his benefit, then the law requires that he pay a reasonable amount for the services or products rendered. Note there was no explicit declaration of intention, but that the intention was implied through the action of the parties. Contracts such as this have also been called contracts "expressed by conduct." This distinguishes them from contracts which are expressed in words.

When a court finds that a contract was intentionally inconsistent, ambiguous, unjust or incomplete, they have the power to abandon various sections and substitute descriptions of what the parties originally intended. A court even has the power of abandoning the entire contract and substituting an agreement which is equitable. When such modifications are made, these agreements are also called "implied contracts."

Quasi Contracts

A **quasi contract** is in reality not a contract at all; it is an action of the court that comes under the name of contract. Let's say that, as a consultant, you have been forced to pay for an item which rightly should have been reimbursed to you by your client. Your client refuses to pay for it and you bring him to court. If the court orders that your client pay for what he rightfully should have paid for in the beginning, this action is referred to as a **quasi contract.** It is obviously not a contract, since one of the parties is not in agreement; in fact, in some situations, one of the parties is in strong disagreement. In an ordinary contract, it is the contract itself which defines and delineates the obligations. In contrast, it is the obligation which defines the quasi contract.

Subcontracts

A **subcontract** arises where one party has agreed to perform a particular piece of work and then contracts a third party to assist him in completing all or part of his obligations which are stated in the original contract. The third party is said to be a **subcontractor** to the original contractor. In some situations, the original contractor is referred to as the **prime contractor,** or just **prime** for short. If, for example, a programmer contracts with a client to deliver a computer program, he may subcontract to another programmer to get all or part of the job accomplished. If you are planning to subcontract some of your work to others, it is wise to keep your client informed of this. In fact, if the contract states that you will be doing the work, you are, of course, legally bound to do it yourself.

CONTRACT BASICS

Should a Contract Be in Writing?

Must a contract be in writing to be enforceable? Though many people believe that they must have a contract in writing, this is incorrect. All agreements which meet the requisite conditions that we've been considering are enforceable, whether or not they are in writing.

Of course, from a good business point of view, you should always try to get a contract or statement of work, no matter how small the job. However, from a legal point of view, it is not always necessary to have a contract in writing. Some situations where you must have your contract in writing are as follows:

1. Contracts which require more than one year to perform. This is important in engineering and data processing because a large number of these contracts are long in duration.
2. Contracts which include the sale of interest in land, or for the sale of land itself.
3. Contracts in which one person promises to pay the debts of someone else.

Contractual Intention

If a contract is written correctly, it can be enforced according to its terms. One of the first things a court will do is to examine the contract and determine what the parties originally intended. Once the court decides what the parties intended, the court will then decide if what the parties intended was lawful. If it is found that one of the parties had intentions that greatly differ from what the contract states, then the court must determine whether the contract is indeed effectual. After all, a meeting of the minds should have taken place.

When the court attempts to obtain a meaning of the contract, the court tries to obtain the originally intended meaning of every word. Often the parties have come to court because of ambiguities in the original terminology. Ambiguities might lie in a particular word or set of words, and the court tries to determine the way in which these words were originally intended. First, the court will attach regular meanings to the words. These are the meanings that one would normally find in a regular dictionary. In engineering and data processing there will often be technical terms used in the contract. The courts will accept the trade definitions if both parties to the contract can show that they used them in the traditional sense.

When you draw up a contract as a consultant, you should make sure that the technical words you employ are explicitly defined in the contract. There are other areas where a consultant should be careful. For example, when quantities are expressed in engineering terms, these quantities often have associated units. You should be consistent with your units and try not to mix them; errors are more probable in such cases.

Contracts Considered as a Whole Unit

Divisible contracts are enforceable agreements which consist of two or more parts. These parts call for particular actions by the parties involved. If one section of the contract is in conflict with another section, the court will usually give the more specific and important section of the contract the greater consideration. However, it should be emphasized that when the court looks at the contract, it first considers the contract as a whole, and then will investigate the various sections. The contract must be understood as a whole, and this even holds when a contract is partly oral and partly written. Of course, if the contract falls under the parole evidence rule, then any oral agreements that occur prior to the written portion of the contract might be excluded.

When the court is trying to determine the intent of the contract as a whole, it may include in its investigation: letters, memoranda, notebooks, and anything else that is considered by one or both parties as an item that led to the creation of the original document. Sometimes all the items combined with the original contract are considered as a whole in order to determine the original intent of the parties.

If You Draft a Contract

Frequently the consultant has standard consulting forms of his own or he's asked by the client to draft a contract or a statement of work. The consultant should understand that when the courts interpret the contract, the interpretation is often in favor of the party who did not draft the original document. Thus, if there are two reasonable interpretations that can be drawn from a contract, it's possible that the court could decide in favor of your client due to the fact that you originally wrote the document. Although it is always to your advantage to write as much of the contract as possible, it is also, of course, advantageous to avoid a court's interpreting a document against your favor. One way that you can minimize such problem is to use standard forms of agreement such as those that are published by the National Society of Professional Engineers. These documents have been written by professionals and act, in a sense, as a third party document. They have been carefully written so that they are fair to both parties, and if you can successfully use documents of this type, you could minimize many interpretation problems that could arise in court. See the Appendix to review sample contracts.

The terms of a contract should be written in as straightforward a fashion as possible. No important terms should be left out of the contract. If important items are left out, the parties must try to employ their faulty memories to interpret what they originally meant. Often, it is differences such as this that can initiate court action.

You'll also be faced with clients who will want to use their particular client agreement forms. In such cases, you should take the agreement forms back to your office and carefully go over them in every detail. If you have any question with regard to their form or content, you should go over them with your attorney. If there is anything in the original form with which you do not agree, make sure that this section of the contract is either rewritten or is struck from the original document. Also, don't hesitate to add any comments or other documentation to the contract form they give you. The more data that you supply to support your position, the safer you will be. The court will put great weight on any supporting items that you have added to the contract.

Add Protective Provisions

To protect yourself, you should make a list of every possible item that could possibly impede your attainment of your contractual objectives and then try to include a clause for each one. If your client objects, you'll have to go over each situation with him and come to a fair agreement. However, if you don't try to include them, you are putting yourself into a situation in which you are vulnerable.

If you are about to take on work that could possibly be affected by strikes or labor trouble, you should definitely have a clause within the contract which protects you. Likewise, if there is any doubt that you will be able to receive the materials that are necessary for your execution of your obligations, you should include a clause which states that you are free from your obligations if you cannot obtain such materials. There are probably hundreds of similar examples. The concept to remember is: if you're in doubt, write a clause about it.

Out-of-State Laws

The consultant might find that his work is called for in states other than the one in which he resides. Since there exist fifty court systems with different statutes in each, it might not be clear to the consultant as to which state would apply or which court would have jurisdiction.

The consultant should distinguish between:

1. The state in which the contract is established.
2. The state or states in which the parties have their permanent home or office.
3. The state in which the contract is to be executed.
4. Any state which is mentioned in the contract which could be the state whose law is to govern the conditions of the contract. The consultant should take notice if a contract states that it is to be governed by the laws of a particular state other than the states in which the consultant is registered.

When you are consulting out of state, you should bear in mind that the state in which the contract is made is not necessarily the state in which a client company resides. The state that governs the contract is sometimes determined by where the last essential act to the creation of the contract was performed. Let us assume that you have a consulting practice which operates in California. You negotiate with a client in his offices in New York State and take home an unsigned contract for review. After reviewing the contract, you find it satisfactory and sign it while in your California office. When you mail it, the acceptance becomes effective, and this corresponds to the last act in the formation of a contract. Thus, in this situation, the laws of the State of California would govern the contract. Of course, if the contract were written such that it said in the contract that the State of New York or some other state law was to govern the terms of the contract, then that statement would take precedence.

If one or both of the contracting parties is a corporation, then the capacity in which the corporation is to perform is determined by the law of the state of incorporation. Of course, the state in which the contract was initially established still determines whether or not the contract is valid and enforceable. In many circumstances matters related to the performance of the contract are often determined by the laws of the state in which the contract is to be carried out. Court rulings have indicated that where many states are involved, the state is chosen which has the most involvement with the contract. This, of course, will put heavy emphasis on the state in which the contract was originally established, and where the performance of the contract is to be carried out.

PROFESSIONAL LICENSING AND CONTRACTS

In order for a contract to be valid, the contract should not contain:

1. Statements that are contrary to public policy.
2. Statements that are contrary to common or statute law.
3. Anything that violates human rights.
4. Statements that are fraudulent.

The contract would also be deemed illegal if the contract were made under duress or undue influence.

In technical consulting there are examples of contracts that are illegal or become void since one of the parties has not met the requirements of the state professional licensing and registration laws. For many types of professional technical work, state regulations require that the consultant be a registered professional. Even if the consultant who was a party to the contract has satisfactorily met the registration and licensing requirements of his state of residency, he could still be working with a void contract if he is not registered in the states in which the contract is to be carried out. If a contract specifically states that it is to be governed by the laws of a state other than that in which the consultant is registered, then the contract could be void. Also, if the contract states that it is to be governed by the laws of another state, and this is not the state in which the consultant's design or products are to be employed, then there's a possibility that the consultant must be registered in both states, since by implication, he's practicing in both localities. When a consultant has questions of this sort, he should employ the use of an attorney or obtain information from a professional technical society. Such societies operate on a state and a national basis, and possess a considerable amount of information which is vital for a proper operation of a consultant's business. These professional societies often have sample contracts which contain sections and provisos for this type of problem.

If a consultant fails to register with the state or forgets to renew his professional engineering license, he will be found to be practicing illegally. If he is working on a contract without a license or with an expired license, he could have difficulty obtaining compensation for his services. Of course, the question of whether he's registered or not would probably not come up until something goes wrong and he goes to court. At that time, he might be required to show his license and he could wind up losing a case solely because it was out of date.

KEEPING YOUR CLIENT OUT OF COURT

Your client might ask you to help him settle a dispute he has with someone else. It should be very clear to even the most inexperienced technical professional that all precautions should be taken in order to prevent having to use a court to settle a dispute. As a consultant, you could find yourself settling many minor disputes that occur between your client and others. For very small disputes your technical opinion will often help. For larger disputes, however, your authority is not usually sufficient, and in the promotion of fairness, your authority should be limited to the more technical aspects of the work that must be performed. However, when it comes to handling legal problems such as contract interpretation, or even the handling of the breach of contracts, you could find yourself in a very sticky position by trying to settle disputes that you are not qualified to settle.

Unless you are an attorney or a professional arbitrator, stay clear of large legal hassles that aren't your concern.

KEEPING YOURSELF OUT OF COURT

A very serious problem arises for a self-employed consultant, or even a consultant with a large company, who recognizes that our court system is in many areas effectively broken down. It is common today to find a situation in which a suit is filed but no settlement is reached for a year or two. Some consulting firms could be in such a situation that by the time they got to court, they could have already gone broke. There is an excellent means of obtaining an early and satisfactory settlement dispute other than the court system. This is the system of professional arbitration. In what follows, I'll give a brief description of the arbitration procedure. In situations where you do not want to wait the unreasonable time it will take for a court to hear your case, you should prepare yourself in advance by providing a section of the contract which stipulates that a form of arbitration should take place in the case of any disputes which may arise as a result of the parties trying to fulfill their obligations.

Arbitration

Arbitration has been defined as follows:

> "Arbitration is an arrangement for taking and abiding by the judgment of selected persons in some disputed matter, instead of carrying it to the established tribunals of justice; and is intended to avoid the formalities, the delay, the expense and vexation of ordinary litigation."[1]

It can be seen from this definition that arbitration is simply the use of a third party or parties who will help settle any dispute that may arise. The purpose of bringing in this third party is to avoid court action and all the hassles that would go along with a serious legal dispute. Throughout the country there exists professional arbitrators, and more information can be obtained from the American Arbitration Association (AAA). For people consulting in the construction industry, the AAA prints a document which could be of assistance entitled "CONSTRUCTION INDUSTRY ARBITRATION RULES." The AAA will send you a copy of this if you will write them at AAA, 140 W. 51st St., N.Y., N.Y. 10020.

What Does an Arbitrator Resolve?

The type of dispute that an arbitrator will try to resolve will differ from case to case. However, since we're dealing with technical consulting here, we can list a few items that commonly fall under an arbitrator's jurisdiction. One of the things that they can help decide is a proper identification and interpretation of items of a technical nature. An arbitrator can also determine what the obligations of the parties are if a contract has not been explicitly written. Financial claims and financial adjustments are also handled by an arbitration tribunal. Arbitrators are also used to determine the level of allowances that might have to be granted in specific situations, as

---

[1] Wauregan Mills Inc. *v.* Textile Workers Union of America, AFL-CIO, 21 Conn. Sup. 134, 146 A. 2d 592, 595 (Supr. Ct. 1958).

well as to settle any claims that develop when schedule problems arise and disputes are initiated over what constitutes a delay and what losses are incurred due to certain delays.

Obviously, arbitrators brought into a situation can only have an authority that is as great as that granted by the individuals who are in dispute. However, if a court grants authority to an arbitrator, then the matter will be different.

There are some matters that cannot be handled by arbitration tribunals. For example, arbitrators are not allowed to decide on matters which are illegal in content. They are allowed to make decisions only where the disputes are disputes of fact or disputes of law.

Courts often recognize the decisions that are made by arbitration tribunals and, in fact the federal and state laws encourage the use of the arbitration system. Examples of modern arbitration statutes are the United States Arbitration Act and the New York Arbitration Law. These statutes recognize the value of arbitration clauses and recognize an arbitration clause as a legal part of a contract. Throughout the construction industry there exists many standard construction, architectural and consulting engineering contracts which employ arbitration clauses. Examples of such standard forms are those of agreement between engineer and architect, and between engineer and associate engineer, both of which are issued by the National Society of Professional Engineers (2029 K Street N.W., Washington, D.C. 20006).

The arbitration laws are there for your use and protection, and should not be considered lightly. Modern day arbitration laws have empowered the courts to uphold arbitration agreements if the court sees fit to do so.

Who Pays for the Arbitration?

Generally speaking, both sides pay for the arbitration costs. It would not be fair to one side if it paid all the costs. This could initiate a problem such that the other side would call for an arbitrator at points where it was not necessary. If both sides share the expenses, then the probability of calling in an arbitrator when it's not necessary is considerably lowered. One of the items that can be included in the arbitration clause is a method of payment for costs that are incurred in arbitration. If this is placed in the clause, then it will minimize one more dispute that could arise at a later time.

The American Arbitration Association has fixed daily rates for their services. In the case of other arbitrators, the amount that is paid varies from service to service depending on the type of work involved and the level of complexity that has to be handled. No set rule regarding rates has been determined, nor is it likely that arbitration rates that apply to all cases will be prepared.

INCLUDE AN ARBITRATION CLAUSE

Whether you expect problems in your consulting practice or not, it is always safe to include an arbitration clause in your contract. Arbitration clauses are recognized under Federal and State laws and they have been responsible for saving untold amounts of money, time and emotional upsets. An example of a construction contract arbitration clause which has been taken from the AAA rules is:

> "Any controversy or claim arising out of or relating to this contract or breach thereof, shall be settled by arbitration in accordance with the Construction Industry Arbitration Rules of the American Arbitration Association, and judgment upon the award rendered by the arbitrator(s) may be entered in any court having jurisdiction."

One of the objectives in preparing an arbitration clause should be to have a clause that will give the arbitrators the necessary power to make a binding decision. Of course, if either party does not agree with the binding decision, then you can always resort to the courts and, if necessary, even appeal the court's decision. If you have a clause which will get you and your client to sit down and settle your disagreements, the chances are very high that you'll be able to reach a decision that is satisfactory to both parties and avoids the courts.

An arbitration clause should give a statement which is similar to the AAA quote above. In addition, the clause should also state a method for selecting an arbitrator. If no method is chosen, then the clause can state that either party may apply to a court of law for a designation of an arbitrator.

Another item which would be useful to include in an arbitration clause is a maximum claim that could be settled through arbitration. Clearly, if the claims become too large, it will be beneficial to work through the court system even though the court system is not as responsive as one would hope for. Although some arbitration clauses contain a section which states that the decision of the arbitrators will be final and not subject to appeal, if one of the parties wishes to take the matter to court afterwards, they may do so.

Many consultants are not in construction-related work, and the arbitration rules that have been set up for the construction industry might not apply. If you are designing your own contract with your attorney, then you could save a considerable amount of time and money by writing to the Professional Associations that are related to your work and determine if they have standard contract forms. Examples of such associations are the National Society of Professional Engineers or the American Consulting Engineer's Council. By writing to associations that are closely affiliated with your specialty, you might be able to obtain standard forms that have clauses that would pertain more closely to your work. Even if you don't use their contract forms, their contracts might have arbitration clauses which could be of benefit to you and your lawyer.

After determining what is available, you can then sit down with your attorney and draw up clauses that will be most pertinent to your type of work. For an example of an arbitration clause, see section 6.9 of the sample contract in Appendix M.

11
Understanding Your Rights and Obligations

Let us assume that you deliver an excellent service to your client, but you do not meet all the requirements stated in your contract. Is your client still obligated to pay you? What are your rights? Can a client hold you to every nit-picking detail in a statement of work? What are your obligations?

The fact of the matter is that just because you sign a contract with someone does not mean that you must become an indentured servant. You have many rights, and knowledge of these rights will make it easy for you to assume your responsibilities properly. You will also be able to work with the peace of mind that comes from knowing that someone will not be able to take advantage of you.

This chapter addresses your rights and obligations in matters like meeting deadlines, incomplete performance, mistakes, contract changes, fraud, etc. It is not my objective to give you a complete legal explanation for all possible situations, but I have attempted to give you a review of certain essential topics, with the understanding that you can add further information as you become more experienced.

DEADLINES

In most technical contracts the date of completion is stipulated. Sometimes there is a series of dates by which tasks must be accomplished. That is, certain deliverables must be presented to one of the parties at a time that is stipulated in the contract.

In consulting, you often find yourself in a position where your deliverables are critical to your client's obligations. It's possible that your client needs your particular inputs to get his other tasks accomplished. Sometimes a short delay can hold your client back and this could cause him to lose a substantial amount of money. When you're negotiating with your client, you should find out what will happen if you don't deliver what you have promised on time. Don't take a generality for an answer. Try to find out specifically what could happen. You might be surprised to find that your client doesn't really know. When asked, he might have to investigate and find out that the time situation is even more critical than he thought. When the time factor is critical, the time is said to be "of the essence."

When you find yourself in situations where time is of the essence, you should plan very carefully so that you know you will be able to meet the deadline. If you have any doubt that you can meet the deadline, you should list those items which you believe could hold you up and mention those items in the contract, stating that these items could be a source of a time delay. Make it clear to your client that you're very concerned with getting the entire program completed on time and obtain any support from him that could assist you in reaching this objective.

HANDLING MISTAKES

Unilateral Mistakes

One cannot be relieved easily from contractual responsibilities by merely showing that a mistake has been made in what was promised. If there is no ambiguity in a contract, a party that makes a mistake in promising a certain thing is generally held to his contractual obligations. In general, courts do not assist an individual who enters into a contract with poor judgment.

Let's say that a company hires a consulting engineer to do some subcontracting work, and that they ask him to design and build some electrical equipment which they plan to install in a building they are constructing. He designs and builds the equipment and makes all the specifications that they stated in their statement of work. Upon receipt of the equipment, they find that it is not satisfactory because the specifications that they had given the consultant were an imprecise statement of what was needed. In such a situation, the company has made what is referred to as a "unilateral mistake." If they try to recoup their losses by having the consultant pay damages or rectify the error on his own time, the court would not allow it.

Generally speaking, when you sign a contract, it's assumed by the law that you understand its contents. After all, that's one of the purposes of a contract—to assist the parties in reaching a common understanding with regard to their mutual obligations.

On the other hand, there are situations in which unilateral mistakes will void a contract. If you offer to do a consulting job for $2,000 and your prospective client knows that it is in reality a $3,000 task, and you have made a mistake in adding your cost estimates, you then have the right to refrain from meeting your obligations. The law will not force you to make good on a promise if it was made under the circumstance that the other party knew you were making a mistake at the time of the agreement. This should be clear, since a contract is supposed to represent a meeting of the minds.

Mutual Mistakes

When both parties make the same mistake in the facts surrounding the contract, the contract is voidable. This is called a mutual (or "bilateral") mistake of fact. An example of a mutual mistake in consulting is a case in which you contract with a client to supply guidance and information to get a particular job done. Upon engaging in the work, both you and your client realize that the amount of work that has to be done and the information that is required is clearly much more than both of you had originally estimated. In such a situation where you both made a mistake by underestimating the circumstances, the contract that you signed could then be voided.

Other Types of Mistakes

Difference in Interpretation

In cases where the language employed in the contract is subject to two interpretations and each party has interpreted the contract differently, then the courts will usually rule that no contract exists since there never really existed a valid offer or acceptance in the first place, i.e., there never was a meeting of the minds.

Clerical Mistakes

When a mistake is clearly clerical, the mistake will not void the contract. A court has the power to rectify a clerical mistake and thus alter the written document to bring it into accordance with what was originally agreed to.

Mutual Mistakes in Technical Matters

Another type of mistake of fact holds with regard to the technical material in an engineering contract. If there has been a mutual mistake with regard to understanding and identifying some particular technical facts, then the contract can become void. For example, if you contract to design what in effect was a multiphase transformer when your client wanted a single-phase transformer, there is no contract due to the fact that it was a misunderstanding of the subject matter by both parties.

Unilateral mistakes can also be made with regard to the technical subject matter. Make sure your client does understand the technical points in the contract and knows what he's getting. If you promise to deliver a particular technical item knowing that it's not what your client expected, he's not obligated to pay you, because there was no real contract in the first place.

Contract Reformation

Suppose you enter into negotiations with your client to design and build a particular electronics circuit and you determine what the specifications of the instrument are to be. Then, when you write the contract, you employ circuit designs that are incorrect due to a drafting error. If indeed there is positive proof of a drafting error due to the clients inputs, the court will usually decide in your favor. In this case the contract need not be made void, but only changed.

Although the courts would probably rule in your favor and change the contract so that you could execute it as you had originally planned, you should take note that these mistakes could be easily avoided by carefully reading every segment of the contract and making sure that these types of mistakes are not present. Before you sign a contract, you should not only go over all the written work and thoroughly understand it, but you should also make sure that the supporting contractual documentation that describes the technical work is clearly understood by you and your client. It does not take long to review the contract and its documents, especially when compared to the great deal of time that must be spent in rectifying mistakes that could have been avoided.

HANDLING INCOMPLETE PERFORMANCE

Ideally, both parties should meet all aspects of the contract in accordance with its specifications, in which case, of course, the contract is discharged. However, in situations where one party does not meet the stipulations of the contract to the letter, he is still entitled to remuneration when his efforts have substantially benefited the other party.

Over the years, a trend has been established by judicial decision which does not require a complete literal compliance with the contract terms. The trend has been towards requiring what is referred to as "substantial performance." Substantial performance occurs when the parties have attained those things that are essential to meeting the objectives and purpose of the contract, even though there may be deviations from the original contract specifications. When one party has substantially performed, the court will protect his rights.

This does not mean the court will protect him from any damages that he might have caused to another by not delivering all that was called for in the contract. The other party who did not receive everything he asked for in the original contract is entitled to some form of recoupment. If you had a contract with your client to deliver certain documents that meet a particular specification, and if you delivered 90% of the drawings that you were supposed to deliver, with 10% of the drawings not meeting the contract specification, the court would probably find that your performance was substantial. However, if the client could prove that he lost $5,000 because you did not meet the contract in full, you would be entitled to the full receipt of the contract consideration *minus* this loss of $5,000 suffered by the client.

An example of this is the case of Boggs, *et al. v.* Shaddurn (65) GA. App. 683 16 S.E. 2d 234, 235 (1941). The builder sued the owner after the owner refused to pay the builder for a house that the builder constructed. The owner claimed that the builder had not met all aspects of the contract. For example, the builder had not waterproofed the basement or graded the yard. The court said:

> "Where a builder has in good faith intended to comply with a contract and has substantially complied with it, although there may be slight defects caused by a

misconstruction of terms of the contract, and the house as built has been received by the owner and is reasonably suited for the purposes intended, the contractor may recover the contract price less the damage on account for such defects. In such a case the true measure of damages is the difference between the value of the house as finished and the house as it ought to have been finished under the contract, plans, and specifications."

The concept of substantial performance is very straightforward. However, in practice, it's difficult to determine what constitutes substantial performance in every situation. The consultant should never think that the concept of substantial performance allows him to freely stray from his obligations. If you deviate from a contract to a substantial degree, a court would rule against you and you would not be able to recover any money owed to you at all. The key to understanding substantial performance in technical consulting is that you should be able to show that any deviation from the contract was unintentional and that the work that you delivered was high-quality workmanship applied to the specific problem at hand.

Your Client's Acceptance of Deliverables

If you take on a consulting job and you deliver materials that are incomplete, the response of your client is extremely important at the time he receives the incomplete or defective deliverables. If he accepts your work where the lack of meeting the specifications is obvious, and it is clear that he has knowledge of any imperfections that are present, then he becomes obligated to pay you. If he promptly objects to the nature of your work, he is not obligated to pay you until a settlement is reached. However, if the imperfections are not obvious and he does not have knowledge of them, this is a much different situation. When you deliver work that is deficient and your client does not understand the degree of deficiency or the deficiencies aren't obvious, he is certainly entitled to sue for a recoupment of losses when he realizes what has happened.

CONTRACT TERMINATION

Contractual Discharge by Agreement

There are a number of legally acceptable ways in which a contract can be terminated. They are accomplished by initiating a provision within the contract or by a subsequent agreement by the parties. Examples are as follows:

Substitution

The client and consultant may decide that another contract would be more satisfactory. In such a case they write a new contract which discharges the old contract; this is called substitution.

Waiver

A waiver is an intentional giving up of a claim. In consulting, a waiver can occur when one party does not demand performance of the other party. For example, if a consultant fails to meet his obligations, and the client does not demand that the stipulations of the contract be met, then we say that those parts of the contract are discharged by waiver. Another example of waiver occurs in the case that the client knows that the consultant is not performing but does not object.

Contractual Provision

If there is a termination clause within the consulting contract and the conditions for that clause are met, then the contract can be terminated on that basis.

Agreement of the Parties

The consultant and his client may simply terminate the contract by mutual agreement. This is also called a rescinding of the contract. When both parties agree, it is a mutual rescision.

Accord and Satisfaction

The parties might change their minds as to what constitutes a satisfactory performance of the contract. If they enter into an agreement which details a new set of performance specifications, they have reached an accord and satisfaction on the first contract. This new agreement is referred to as an accord.

Discharge Due to Acts of the Other Party

If you contract to do a particular task and you are prevented from completing your contract due to acts performed by your client, you could be legally discharged from your contractual responsibilities. An example of this is the case in which your client was supposed to deliver certain information, material, or equipment to you as a necessary prerequisite to getting the job accomplished. If he did not deliver what he promised to deliver, you would no longer have to meet the promises that were made in your original agreement. Note that his breach has to be a substantial break from the original contract, and that a minor breach would not constitute sufficient terms for you to be discharged from your responsibilities.

Discharge by Impossibility

If one of the parties finds that it is impossible to execute his obligations, then there exists a possibility of a contract discharge by impossibility. However, what constitutes an impossibility is up to the court. Even events as unexpected and severe as floods and tornadoes are not usually sufficient conditions to end a contract through discharge by impossibility. In construction, extreme shortages of materials or problems with labor rarely result in successful claims of impossibility.

However, in technical consulting it sometimes does happen that a job is taken on in which it is found that according to the laws of nature or the laws of physics the work cannot be completed to specification. If you find yourself in such a position, the chances are that your client will probably understand and you will be able to rescind the contract by agreement. However, if your client did not go along with this, it would be up to you to show that the contract was indeed impossible to fulfill.

Objective Impossibility

When it is found that it is extremely impractical or impossible to do the work for technical reasons, then the party or parties to the contract might be excused from their obligations. It might occur, however, that the tasks that have to be accomplished were possible at the time the contract was written, but then because situations and circumstances changed the task became impossible.

There doesn't seem to be a great deal of agreement amongst court decisions on how to decide whether a set of circumstances meet the requirements of objective impossibility. Obviously, if you try to claim a release from your obligations on the basis of objective impossibility where the work has not been done due to some negligence on your part, the courts will not grant your request. Usually, an unforeseen happening must occur which prevents you from doing your job. As an example, let's examine the case in which you contract with a client to install some highly technical and sophisticated test equipment in one of his laboratories. Then a fire in one of the laboratories destroys the laboratory to the point that it would not be possible to place your equipment in it. However, your client wants you to place the equipment in another building which you find would not be satisfactory for the performance of the equipment. This being the case, you would find it impossible to meet your responsibilities as you originally believed possible when you drew the contract. This is a case of objective impossibility because you cannot place the equipment in the room that was originally set up for this type of equipment.

There are some decisions from the courts that take the point of view that even though things like acts of nature might intervene, you are still responsible for a performance that you promised and that you are not excused by any impossibility brought on by unforeseeable circumstances. The rule of thumb in cases like these seems to be that if the impossibility occurs as a result of damages to or changes in an area for which your client was supposed to be responsible, then you will usually be discharged of your obligations. In any case, it is wise to write your contracts such that they contain clauses which handle any problems that could arise from unforeseen circumstances. Whenever acts of nature or other unforeseen circumstances appear, the court is placed in the position of deciding which of two innocent parties must sustain the loss. Things being equal, the courts would most likely find that the loss is to be placed on your shoulders, if you were the party who left extra provisions out of the contract which could have protected you.

Subjective Impossibility

Subjective impossibility occurs when one of the parties cannot meet his obligation due to his own inabilities. A subjective impossibility could result if you took on a well-defined job, but when you attempted to do it, you found that you were not really qualified to complete the job correctly. In such situations, the courts will not allow you to discharge your responsibility to your client simply on the basis that you are not qualified to complete your contract. If you take on a job for which you are not qualified, you are still responsible to see that the job gets done. If you are not sure you can do a task for a client, you should tell him so. You might think about contracting with him on the basis that you'll deliver your best effort.

BREACH OF CONTRACT

A breach of contract exists whenever one or both parties to a contract fails to perform the contract without sufficient excuse or justification. As I have already indicated, there are circumstances where one party can enter into a nonperformance and such actions will not amount to an actual breach; i.e., there are a number of valid excuses for non-performance. However, there are also many ways that a breach can come into existence. If one party simply renounces his willingness to perform the duties, a breach will have been committed. Also, if that party acts in such a way that he has made it impossible for himself to fulfill the requirements of the contract, then this will also lead to a breach. Another way a breach can arise is if one party to the contract hinders the other party's performance to the degree that the terms of the contract are put in serious jeopardy.

When a breach has occurred, this does not mean that the contract has ended. When one party does not meet his obligations, the injured party has the option to keep the contract or discharge the contract. In fact, in most cases, the injured party insists that the defaulting party perform actions to meet his obligations. The injured party will most probably resort to legal remedies if his request is denied by the defaulting party.

There are various degrees of breach. A breach of part of the contract is not considered to be a breach of the whole contract if the whole contract is divisible into parts. If minor terms in the contract are not met, these minor breaches will not discharge or cause support for a discharge of a contract. When a minor part of the contract is not met, the contract can still remain valid, and the defaulting party may be liable for any damages he caused by not meeting the terms of the agreement.

In situations where one of the parties is guilty of a total breach, the injured party has the option to continue or stop performing his duties and obligations. Of course, the injured party can then sue to recoup any losses that were incurred.

Anticipatory Breach

When the defaulting party to a contract states in advance of the time for performance that he will not meet his obligations, the injured party has three courses of action that he may follow: (1) he may accept a statement by the defaulting party as a breach of the contract and rescind the contract; (2) he can ignore the defaulting statement and insist that the other party perform in accordance with the terms of the contract; or (3) he may also accept a statement as an anticipatory breach and sue the defaulting party for any losses that have been incurred. Anticipatory breach occurs when one of the parties to a contract announces before his performance date is due that he is either unable or unwilling to fulfill his obligations. In doing this, his guilt is admitted, and this immediately relieves the injured party from any obligations that that party had in the contract. An anticipatory break does not occur when one of the parties threatens to abandon the contract; he must make a positive statement that he will not perform his duties. If one of the parties goes bankrupt and cannot meet his obligations of the contract, it is considered by the courts to be a form of anticipatory breach.

After the defaulting party has announced that he will not meet his obligations, the injured party cannot continue to carry out his obligations and create bills which would increase his damages. If he is to sue for losses, he may sue only for losses which are the result of the breach at the time it is announced.

Remedies for Breach

The injured party has a choice of three alternatives in his attempt to remedy the breach. These three methods of remedy are referred to as: (1) damages, (2) specific performance, and (3) restitution.

Damages

When the injured party seeks to remedy the breach by recovering a sum of money that would place him in a position equivalent to that which he would have been in if the contract had been performed, he is using the method of remedy referred to as "compensatory damages." Though it is not always easy for the court to determine the actual size of the damages that have occurred, the court uses the general rule of determining what it would take to place the injured party into a position that he would have occupied if the defaulting party had actually made good on his contract.

In this case the injured party does not try to obtain compensation in the nature of punitive damages; he merely tries to gain his original objectives. For example, in construction situations when the contractor defaults, the owner generally attempts to recover the cost of having his incomplete job made complete.

Specific Performance

In some situations the injured party may seek remedy by having the courts force the defaulting party to carry out his terms of the contract. This is called the remedy of specific performance. This remedy is seldom resorted to by the courts; in most cases if the court can find another legal remedy which will be adequate, it will use that remedy rather than the remedy of specific performance. The courts will not apply specific performance in situations where it implies an undue hardship or an injustice on the defaulting party. Another situation in which the court might not apply the remedy of specific performance is where the court would find it difficult to supervise the performance of the necessary actions.

When technical contracts are breached, it is found that the courts usually do not employ the remedy of specific performance. For most situations of breach of technical contracts, the remedy of damages is used, if for no other reason than that it would be difficult for a court to supervise such a contract to make sure that the defaulting party met his obligations.

Restitution

If the injured party decides that the only money he would like in return is what was spent during the performance of the contract, he can then file suit for restitution. By seeking a remedy of restitution, he is basically trying to recoup any money that was lost from the time the contract was initiated until the time of the breach.

MISREPRESENTATION

Suppose that you enter into a contract with a client and later find that there exists some false statements that were made innocently by both parties, and these facts have now put you into a disadvantageous position. Could you then claim that the contract was void since you were misled by a false statement? As in many of these legal issues, this area is not altogether black and white. In many situations the court would hold that the contract would remain in force, though where the case is very obvious, the law would protect you and permit you to legally discharge your obligations. In any case, you should remember that in such cases of misrepresentation, where there exists false statements that were entered innocently, you might have a very difficult time winning your case. To avoid such possibilities, it is best to clearly understand the contract and know exactly what is expected of you.

Fraud and False Statements

Fraud exists every day in technical consulting, and you had better know what it is so you can identify it. There is a case of fraud when the following four conditions exist. (1) A party misrepresents a material fact which is (2) known to him to be untrue or made with reckless indifference as to whether it is true, (3) with the intention of causing the other party to enter into a contract, and (4) leading to the detriment of the other party. Fraud in technical consulting arises most commonly when a consulting firm tries to win a contract by misstating their abilities in order to mislead the client. They then produce a product that does not meet the specification of the contract.

In many circumstances the client does not sue to recoup his losses. This is because in most technical work it is difficult to prove fraud, since it is difficult to prove that one of the parties gave an intentional misstatement of fact in order to win the contract. Just because many of these cases don't wind up in court gives you no excuse for keeping your eyes closed to such situations. It is fraud when anyone intentionally misstates a fact in order to deceitfully win a contract to the detriment of the client. Be able to recognize it and stay away from individuals and companies who think nothing of doing business in this fashion—who believe that this is "how the game is played."

The success of your company and your future will be dependent on the trust that you can generate with your clients. Because this trust starts at the bargaining table, you should never make misstatements of fact with intentions to deceive.

When Is Expert Advice Fraudulent?

In general, statements such as "our product is the best on the market" cannot be considered fraudulent, even when obviously false, because they are matters of opinion. Likewise, in the ordinary course of consulting affairs, a misstatement of opinion or a remark about the value of something is not regarded as fraudulent since the client receiving the information can and should recognize that it is merely the consultant's opinion. However, when an expert states that he has certain abilities to complete a particular job, he's making a statement of fact. If he falsely claims that he has done things in the past in order to support himself in the winning of a contract, he is setting up a transaction between the parties which can lead to fraud.

Simply stated, if your client has the erroneous impression that certain things are true because of your statements, and you know that these facts are not true, you commit fraud if you do not disclose the truth before your client takes the incorrect information and makes decisions which damage his position. Remember, we're talking about facts here and not opinions. When your client asks you for facts, and you don't have the requested information, you should indicate it. When he asks you for your opinion, you can give him such, noting clearly to him that it is your opinion. Don't ever give your opinion and make it sound as if it is fact.

Half Truths

It's also helpful to remember that a partial truth which has the same effect of misleading your client can also lead to fraud. Such a half truth can be considered by the courts to be similar to a statement that was entirely false.

Unintentional Misinformation

Another circumstance arises in consulting where you may have inadvertently given your client incorrect information, and then at a later time you find out that you had made a mistake. It is your obligation to immediately contact your client and indicate that some of the information you passed along to him was incorrect. If you do not take that responsibility, in effect, you are intending to have him remain working with incorrect data. This, again, could lead to fraud.

Statements Which Prevent the Client from Making Further Inquiries

Another type of fraud that can be committed by a consultant is to make a statement which prevents the client from inquiring further into an area where he should be obtaining more information. An example is the case in which you are asked for particular facts, and the proper answer might be that you don't have any information on the topic in question. This answer could initiate a continuing inquiry by the client so that he eventually obtained the correct information from someone else. If, however, you give him a partial truth which prevents him from the continuation of a proper inquiry, then this could also lead to fraud.

Remedies to Fraud

The deceived party in a contract can allow the contract to stand, and if he chooses, may sue to recoup what he has lost as a result of the deceit. He also has the option to rescind the contract, though this right can be lost by the injured party if he knows about the misrepresentation but does not do anything about it within a reasonable time. If he does not decide to rescind the contract, the contract is valid and remains binding. If the injured party decides to rescind the contract, he can try to recover any payments that were made in the performance of the contract. In no circumstances can the wrongdoer rescind the contract.

12

Delivering Your Service

Up to this point we have minimized the importance of what might seem to be a critical issue in consulting—the actual delivery of the service. It is true that most technical professionals have exaggerated fears that their delivery of their services will be inadequate. It is also the case that anyone who delivers fine results as an employee will likely do the same as a consultant. However, there are several important issues related to service delivery that should be kept in mind. The way you deliver your product or service as a consultant will determine how much repeat business you will receive, and thus how successful you will be.

DELIVERING WHAT YOUR CLIENT REALLY NEEDS

Know Your Client's Objectives

At a recent seminar on consulting I asked a large number of attendees what they delivered to their clients. What were they paid for? I received answers like: "schematics, reports, mechanical drawings, solutions to problems, etc." These items are often what you deliver, and they are part of the reason you get paid. However, if all you deliver is a report or a drawing, you will never get repeat business consistently.

I had a client not too long ago who called me in to bring his department up to speed on some optical filters I had designed. They said that their business objectives were to train some of their researchers in optical filters so they could remain competitive in their area of expertise. After I was consulting to them for a week or two I found out through consistent probing that my client had a very strong personal urge to convince the government to fund an area of research that he felt was critical to the nation's defense. It was a personal challenge for him to convince the government to move in the direction that he knew was best. I helped him reach this objective as well, and obtained more repeat business.

Deliver More Than a Report

From that point on, I not only brought his department up to speed as promised, but I wrote many one or two page reports that showed my client how the information could be used to convince the government that his approach was a valid one. I spent hours showing him *how my information could be used* to help him convince the government that his approach was best.

This is in contrast to what an inexperienced consultant would have done. I could have written a long and detailed report, dropped it on my client's desk and billed him. That's not delivering your product—that is just dropping a report on someone's desk!

To deliver your product or service correctly, you must know your client's objectives. Your principal service is helping him meet these objectives—that's what you are really getting paid for. Help your client reach his objectives and you will get repeat business.

This might seem like common sense, but whenever I ask consultants what their principal service has been, I always get the same type of answer: "reports, designs, drawings, solutions . . . ." It's very important to realize that once you have solved an important problem for a client, he must be educated as to how he is to use the solution to help him reach his objectives.

Where Consultants Fail

As a consultant you will sometimes be hired to interpret or evaluate the work of another consultant. After reviewing the work, you might find that the other consultant wrote a report that only an expert in the field could understand. Instead of explaining what the results meant, he just wrote a fancy report and billed the client. Did he help his client reach an objective? No. He confused his client, and now the client thinks he needs an independent evaluation. You are not being brought in for an independent appraisal; you're being brought in to help pull this client out of the mud. He's confused and upset. He just got through paying a consultant several thousand dollars to come in and confuse the situation.

Your Client Will Not Always Know How to Utilize Your Skills

Your client will not force you to sit down and explain your results to him. You must take the initiative and review your results with your client and make sure your client understands all the implications of your endeavors.

If possible, try to avoid writing long reports. I get much better results by writing a few typed pages and then reviewing just a few results with my client. I've noticed that an overworked client often has no desire to read a long report. I had one client who never had a chance to read. He was either in a meeting or on the phone. He was so busy that one of the few times he read was when he went to the men's room. I started writing a large number of very short reports and he had time to read them. If he had the choice between picking up one of my one-page reports or someone else's fifty-page manuscript, he would gladly choose mine, because he knew he could finish it before his next interruption.

GENERAL HINTS IN DELIVERING YOUR SERVICE

A Little Known Secret

There's a fact about consulting that may come as a surprise to some of you. It's a simple rule that guides a consultant to success. The rule is this: *While consulting, do everything in your power to work yourself out of a job by making your client as self-sufficient and independent as possible.*

What's this? Work yourself out of a job? Thought we were trying to make money—not lose it! This one idea, practiced diligently, has guided me to success. Let me explain how it works.

Let's take a very typical situation. A new client brings you in as a consultant after he's tried in vain to hire someone who is knowledgeable in your field. Since he's behind in his project, he's delighted to have you aboard. If you wish, you could stay with him and his project for quite a while. Sounds pretty good, doesn't it? It happens all the time.

But wait a minute, you're supposed to be a consultant, and you're supposed to help your client meet his goals. *What was his original goal?* To have a full time employee who could support his project and be an asset to the company. If you're a sincere consultant, your priority will be to first solve your client's most pressing technical problems and then help your client get a full-time employee to replace you.

Why do you do such a thing? Because that's what's best for your client. He's paying you good money to help him reach his goals; he's not paying you to stick around for as long as you can. He, like many other businessmen, is plagued with parasites and he doesn't need another one!

So now that you've worked yourself out of a job, what are you to do? Don't worry—you've just made yourself an ally who will never forget you. Think about it. Put yourself in your client's shoes. You came along and helped him out by doing what was best for him. Who do you think he's going to call the next time he's in a bind? The guy he had before you who bled him dry? No way. He's going to think of the person who had the courage to do what was right. *He's going to think of you.* That's what follow-on business in consulting is all about.

Adjusting Your Schedule

If you promote yourself properly, you might wind up with several clients simultaneously, and it will become important not to overload yourself to the degree that the quality of your service drops. If you have deadlines to meet, you should decide whether or not you can handle another client before you sign another contract. If you have any doubts, you should make the necessary adjustments in any new contract you take on, clearly delineating the time you have available.

Consider purchasing a daily scheduling or reminder book. Once you're in business for yourself you'll have to get more control over your time. It also helps to plan the day and follow the plan as best as you can. Make a habit of keeping this scheduling book in your briefcase and making sure you have it with you at all times. If you don't use one of these scheduling books, you might find yourself with appointments to see two clients at the same time.

Coming Up to Speed

Sometimes a consultant must learn new techniques before he can really contribute to solving his client's problem. No matter what consulting contract you work on, there's always a period during which you must learn and understand exactly what the problem is. Before you can contribute to a solution, there is a learning period where you must "come up to speed." For some problems, this learning period can be very long; it can even be a matter of months. On other problems with which you have had more experience, the learning phase might only be a couple of days. No client expects you to walk in the door, listen to the problem, and immediately contribute to its answer.

In my own career, I often take three or four weeks on a job just to become familiar with the various technical and personnel aspects of the problem. Then, after becoming thoroughly familiar with what the job entails, I can then sit down to do a competent job in solving it. I've found that it generally takes between 10% and 20% of the entire job just to understand the problem sufficiently to begin to solve it. You should not feel uncomfortable in requiring a grace period, during which your client is paying you to come up to speed on his problems.

Exposing Your Ignorance

Another fear that some consultants have is that some of their ignorance will be exposed in the process of doing the job. It's true that you should know your fundamentals, but you don't have to feel that just because you're a consultant that you have to be a genius or the world's greatest expert in your area. There's nothing wrong with expressing your ignorance about a particular problem. I have never had a client who ridiculed me for what I didn't know. If you explain your background to him, he won't expect you to be expert in all areas. Your client is paying you to do a good job in a relatively short period of time. That does not mean that you have to be extremely fast in getting the job done, nor that you have to know everything there is to know about the problem. What's important is that you take a sincere attitude and do everything necessary to get the job done properly.

The most important factor in getting the job done is to understand the nature of the problem. In order to do this, you have to ask questions, and you have to expose your ignorance. Problem solving is most rapidly handled by exposing your ignorance of those aspects of the problem where you feel you need help. After all, it's what you don't know that prevents finding a solution. If you feel strange about exposing your ignorance on a problem, you should speak to your client beforehand and tell him exactly what your style is. Explain to him that for you to solve the problem quickly, you're going to have to ask a lot of questions, and at times the questions might seem rather basic. He will appreciate your candor.

I treat every consulting contract as a learning experience, concerning both technical problem solving and handling people. I consider it a personal obligation to learn on the job. If I don't ask questions, I don't learn, and if I don't learn, I'm not expanding my talents.

Also, you'll find in many situations that people within the company itself already have many of the answers. You should not hesitate to go anywhere within your client's company to collect information.

Interacting with New People

As you go from contract to contract, you'll be constantly dealing with a new set of people. This has distinct advantages. Primarily, you will accumulate contacts with a large number of talented people with whom you can potentially work in the future. After you've accumulated a large number of contacts, you can then select the cream of the crop for your clients. This is when consulting really starts to be fun.

By going from client to client, of course, you will have to get used to working with new people. If you're uncomfortable with this, you should enroll in the communication course mentioned in Chapter Fifteen. You will find that your ability to handle a large number of new acquaintances will increase and your fears will subside considerably. Like anything else, communication takes practice.

With each new contract, you have to make a judgment of how much time to spend doing technical work, and how much time to spend socializing. When I first started contracting, I used to concentrate very hard on the technical aspects of the job. When it came time for lunch or doing a little socializing with some of the other technical people in the area, I often refused. I felt I should just get the job done in the shortest time possible. As I gained more experience, I realized that many of the problems that I had to solve required interacting with other individuals within the firm. I quickly found that it was very important that I spent some of my consulting time getting to know them, their problems, and some of their own idiosyncrasies. When I needed them to contribute to solving the problem I was working on, they would be available and would work with me in a cooperative fashion.

Therefore, when you go on a new assignment, don't feel that you have to spend every minute at your desk working at the technical aspects of the problem. Get to know the managers and the employees who are associated with the project. Develop a good rapport with them. Remember that your final product is not just a computer program, a system that runs better, or a final report, but a customer and his employees who understand what you've done and who are using your results to meet their objectives. Part of delivering your product is communicating your results to them. If you're going to communicate your results properly, you have to develop a sufficient rapport with them. You'll be judged not only on how well you solved your problem technically, but also on how well you communicated the results to your client and his employees.

Use Others to Get the Job Done

Thomas Edison, who had over a thousand patents to his credit, was hailed by a writer for the abundance of his inventions. To this he replied, "I am not a great inventor. The only invention I can claim as absolutely original is the phonograph. I'm an awfully good sponge. I absorb ideas from every source I can and put them to practical use, and then I improve them until they become of some value. The ideas I use are mostly the ideas of other people who don't develop them for themselves."

One reason Edison was productive was that his personality was such that he had no fear of using the ideas of others to get a job done. His creative genius was largely in taking other men's ideas and bringing them to the point where something practical could be done with them. He was neither trying to prove to others how creative he was, nor trying to get attention by showing people how original he could be. The credit he received was due to him because he was highly productive; he specialized in getting the job done.

Clients have often told me that they're overloaded with prima donnas and geniuses. What clients are looking for is people who will just get the job done. One of the most efficient ways to get a job done is to *make use of what has been done by others on the project and then employ as many people as possible to get answers to your questions.* Sometimes a consultant believes that he must do everything himself. This is not the case. Use everything you can to get the job done; other people's notes and experience can save you a great deal of time and effort. Sometimes you have to do very little original work yourself, and you just act as a focal point for other people's data and experience. When this is done, you can often get a job done in one-tenth the time it would take you to do it all yourself. It's true you won't come across to some people as a creative genius. But to the person who hired you, you will be considered a gift sent from on High.

Don't Live by the Sword

As a consultant, you will often be brought into a situation in which many mistakes have been made and the project is in deep trouble. You might be asked to give a candid assessment of how bad off the situation really is, because the manager who brought you in does not have the technical experience to be able to judge. After studying the situation, you will have a choice of how you express to him what is right and what is wrong with what you have found.

Almost anyone can go into an area and emphasize what's wrong with it. When this occurs in consulting, it can drive the client into a state of panic and cause him to fly off the handle with his employees. It has been said that fear is a very powerful and motivating force. Because of this, a consultant's opinion can often do more harm than good if rendered without regard for the potential effects it can create.

The important thing to avoid when you're brought into a consulting position is making rash, intimidating statements about errors that have been made on the project. Just get the job done. Most mistakes that you'll find in many engineering and data processing situations are mistakes that are made by good-willed, but mistrained, individuals who did not know how to carry out a job successfully. Sometimes they are the victims of poor management, or they could be working in a time schedule that's unrealistic. To blame them unjustly will help neither you nor your client.

Change Your Life-Style to Create Time for Yourself

To become successful and to remain that way, you must realize that time is a valuable commodity. You're going to have to create more of it for yourself. The way to do this is to start eliminating a large number of tasks that you ordinarily do during the day. You must delegate and have other people do as many tasks for you as possible. Even if you have to pay someone considerably, you should definitely get these tasks off of your plate. This includes going to the store, putting gas or oil in your car, automobile maintenance, handling letters, etc. Your spouse, members of your family, or an employee can start taking on these tasks for you. It's essential to recognize that if you're going to expand, you're going to have to give up your present life-style.

Of course, a side benefit of this recommendation is that you're going to have time to relax. A surprising thing will happen. Those relaxation periods will become moments of business creativity. If you're busy all the time, you won't be at liberty to think of ways to be more efficient. It is also fun to have more time available for other activities. After becoming a consultant, you can usually make somewhere between sixty and one hundred thousand dollars a year. There's no reason why you can't take five or ten percent of that money and invest it in people who will do extra tasks for you. Delegate at every opportunity. Treat yourself to the joy of having people available to take some of the load off your shoulders.

Documentation

A very important habit which helps result in a successful consulting business is the willingness to regularly document your work. Although documenting your ideas may take time which you feel you don't have, in the long run it will save you time and make you money.

Furthermore, you'll increase the value of your service. After you have consulted for a while, you will be asked to solve problems that are similar to problems you have solved in the past. If you have documented your work, you'll be able to refer to that work immediately. Your reponse time to a client is exceedingly important, and if you have data which is already written that you can apply to your present problem, you will be able to give quick, cogent solutions to your clients' problems. This gives you a big advantage over those who cannot do this.

There is also the area of the fixed price contract. Sometimes you'll bid on a fixed-price contract, and part of the contract will contain work that you have already done in the past. The time you will save in not having to reproduce work you have already done will save you tremendous amounts of money. After you've documented most of your work, you'll find that the fixed-price contract is the most lucrative.

Another reason for documentation becomes apparent when you write down your ideas. It's in the writing stage where flaws in one's work are found or weaknesses in the argument become apparent. Writing reports on your results not only clears your own mind, but gives you a better perspective of the problem.

Getting Paid

If you have chosen your client wisely, you'll have no trouble getting paid. If you didn't look into a client's background, or if you weren't concerned with a client's business ethics, you might face the rather unpleasant situation of either not getting paid, or having your check sent to you months after you need it. Aside from this very important tip on choosing your clients, you should also keep in mind certain data on invoicing.

Invoicing

Invoice your client immediately upon completion of your work. Get the invoice into the hands of the proper person without delay. If you've selected your client wisely, you'll receive your check promptly.

Find out from your client before the work is complete, preferably in the planning and negotiating stage, how to invoice him. You don't want to cause your client problems by billing him incorrectly or billing the wrong person in the company. If the project has a project number for billing, make sure you place this number on your invoice.

Letter of Recommendation

Let's assume that your contract has come to an end, you've delivered a good service, and your client is happy with your performance. You're not through. At this point it's important to have your client write a formal letter of recommendation stating what you did and why he was satisfied with your work. After you have four or five letters of recommendation to show to future clients, you will find that getting new contracts is much easier. One of the reasons people hesitate in hiring a new employee or bringing in a new consultant is the fear that this person might not do the job. Having a letter of recommendation, or better yet, four or five of them, relieves most prospective clients of any fears.

There is another aspect of the letter of recommendation that is significant. After you've done a good job for somebody and he's written you a letter of recommendation, he has now publicly acknowledged the fact that

GENERAL HINTS IN DELIVERING YOUR SERVICES

he is on your side and he's pitching for you. Consider what happens if, three or four months later, he's in a meeting and your name comes up as a possible source of assistance. The company is considering bringing you in again, and they turn to your former client and ask him what he thought of you. After writing you a good letter of recommendation, he has already formally stated that he's on your side, and you can count on a positive comment.

Your Next Contract

There's a very good chance you'll get another contract from your present client. However, don't accept another contract until you have reviewed all the new opportunities that you obtain through your promotion program (Chapter Four). Sometimes you'll find that there are clients that are so happy with your work that they'll want you associated with their firm, almost on a full-time basis. There can be a problem with this. Real security and financial strength comes from having a wide base of clients and obtaining considerable exposure within your field of expertise. If you stay with a client too long you will not get the maximum of exposure. Accept those offers that will provide the most challenge and also provide an enjoyable atmosphere. Being in business for yourself should not only be a prosperous activity, but it should also be fun.

If your client wants to give you another contract, but you would rather work on another project, you should consider hiring another person to do the job for you on a subcontract basis.

13

Tax Savings and Financial Planning

There are considerable tax savings available to consultants, and your key to taking ample advantage of these savings lies in understanding a few tax basics. Though you should hire a good accountant to handle the details for you, you should have a firm grasp of tax fundamentals so you can talk productively and intelligently with him. He can save you large sums of money if you are familiar with how the tax laws affect your business. This chapter will give you some of these basics and show you how some of the tax laws are related to consulting, as well as giving you other fundamental financial advice.

The material in this chapter should not be considered the final or most authoritative word on current IRS policy. When it comes to taxes, it is a must that you associate yourself with someone who can properly represent you should the necessity arise.

CHOOSING A FINANCIAL ADVISOR

There are two major steps to monetary success. "First ya gotta earn it" and "then ya gotta keep it." It has been said that you have to be smart to earn it, but a genius to keep it.

As a consultant, you won't have much problem earning the money you need, but keeping it is another story. I know I've made it sound thus far as if the consulting world is a pretty rosy place, and as if there is no one who is "out to get you." Well, let me say here and now that there *are* people out there who are after your money, your success, and anything else they can get their hands on. They are called IRS agents.

The way they are taught to think is quite simple. If you're a loser they will leave you alone. If you're a winner, then they want to make you a loser. Therefore, if you are going to be a winner, you had better have a good accountant.

A Financial Advisor's Value to You

I would like to give you a perspective of how valuable a financial advisor can be by just listing a few topics where he can help you decide what's best for you.

1. Deciding between a sole proprietorship, a partnership or a corporation.
2. Helping you incorporate.
3. Assisting you with paying off debts or collecting them.
4. Helping you increase your credit standing.
5. Automobile depreciation or leasing.
6. Creating and understanding your business' financial statement.
7. Personal and business investments.
8. Insurance.
    a. Partnership policies
    b. Health and medical reimbursement plans
    c. Auto
    d. Casualty
    e. Liability
    f. Life
    g. Business interruption
    h. Group term life
    i. Buy-Sell agreements
9. Leasing versus owning equipment.
10. Retirement plans.
11. New business ventures.
12. Bookkeeping.
13. Real estate.
14. Wills.
15. Taxes (State and Federal).
    a. Depreciation methods and strategy
    b. Investment credits
    c. Sales taxes
    d. Regular write-offs
    e. Tax shelters
    f. Tax planning
16. Buying or selling a business.
17. Forming a family partnership.
18. Estate formation and planning.
19. Trust establishment.
20. Handling capital gains and losses.

Your advisor can help you on all sorts of matters, and one of the advantages of going into business for yourself is that you can afford to start dealing with a real professional with regard to your personal and business finances. It should also be mentioned that your accountant's fees are tax deductible.

There Are Different Types of Accountants

The acronym "CPA" stands for Certified Public Accountant. All accountants are not CPA's. In order to become a CPA, an accountant must become licensed by passing a rather rigorous exam given by the state in which he wishes to practice. This is a three-day exam which tests the accountants' abilities in accounting principles, taxes, business finance, etc. Since each state issues their own license, the requirements differ from state to state. For example, in some states the accountant must have two years of practical experience in matters of taxes, general accounting and auditing.

Although there are about 150,000 CPA's in the United States, less than half of them are in private practice. Many of the CPA's work for what is called "The Big Eight," the largest national firms. Just because someone has been an accountant for twenty years doesn't mean he can be of value to you. If he's worked as a CPA for the Big Eight or for large industrial firms, he might not know a thing about helping a small business make it. You must find an accountant who primarily deals with small business matters.

The accountant that is right for you might not even be a CPA. In some states there is a breed of accountants called "Public Accountants," which should not be confused with CPA's. In some states a PA does not even need a college degree, and they have not had to take the rigorous CPA exam. Some states no longer issue a Public Accountant license.

Many accountants are neither PA's nor CPA's. They are just accountants who usually have a college degree in accounting, but have not passed any state licensing exams. Don't rule out these accountants just because they aren't CPAs. Just as there are many competent engineers who are not Professional Engineers, there are many competent accountants who are not CPA's.

There is still another group of financial advisors that might not come under the classification of a college grad, accountant, or CPA. This is the Enrolled Agent. An Enrolled Agent is anyone who has passed a test given by the Internal Revenue Service. This test is two days long and is rigorous. Those who pass it know their tax law quite well. Enrolled Agents are permitted to represent their clients in tax matters, and some of them know more about tax matters than most CPA's.

At the top of the advisor list is a tax attorney. The tax attorney is the person who can really let you know how much you can get away with. He not only knows the tax rates and laws—he knows how they are interpreted by the courts. After you have worked out your tax strategy with your financial advisor, you might want to have a tax attorney look it over and give you his professional point of view.

Typical rates paid for financial advice in 1982 are as follows:

Tax Attorney	$100-200 per hour
CPA	$40-60 per hour
Enrolled Agent	$25-40 per hour
Tax Preparer	$15-30 per hour
Senior Accountant	$10-20 per hour
Bookkeeper	$10-15 per hour

Decide How Your Are Going to Use Your Accountant

Before you pick an accountant, you should have some idea of how you are going to use him. For example, though my accountant does most of the work on my taxes, I prefer to learn a little bit more on the subject every year so that I don't have to rely on someone else exclusively. For me, relying on anyone else exclusively with regard to taxes is just not my style. If someone grosses $100,000 per year and falls into the 50% tax bracket, then the slightest advantage can mean thousands of dollars saved.

To keep up to speed on tax developments and how they affect the small businessman, I have subscribed to a monthly tax service called "SMALL BUSINESS TAX CONTROL" offered by CAPITOL PUBLICATIONS, INC., SUITE G-12, 2430 PENNSYLVANIA AVENUE, N.W., WASHINGTON, D.C. 20037. The price is $76.00 per year, and this includes postage and handling for their monthly updates. It's a good service and, for me, $76.00 represents just two hours of a CPA's time. Moreover, I have hundreds of pages of tax information at my disposal at all times. When I can't interpret what I'm reading and my accountant's dictionary doesn't pull me out of the mud, I can always call my accountant.

My style is one of wanting to play this tax game to the hilt. I think taxing anyone at 40% to 50% is insane, and I like beating the IRS at their game using their rules. Now that's not everyone's style. A consultant friend of mine is grossing $75,000 this year, and he could care less about tax write-offs or the like. He apparently doesn't mind giving Uncle Sam half his income. So the way he will use his accountant is quite different than the way I use mine.

I have my bookkeeping arranged so that at the end of the year I can rapidly start on my tax forms. After I've found all the deductions possible, I call in my accountant to look for more. Between the two of us we find and use every loophole possible.

How Much Will Your Accountant Cost You?

He shouldn't cost you a dime. He should save and make you money. Don't fall into the trap of looking at an accountant from the point of view of $30, $40, or $50 per hour. Sure you are going to spend money, but you must look at the entire exchange.

CHOOSING A FINANCIAL ADVISOR

Let's say your gross is $70,000 this year. Your accountant could easily save you $3,000 per year in taxes and another $2,000 per year in estate planning. So if he charges you $500.00 for a $5,000 savings, don't look at it as a $500 cost. It's a $4,500 savings.

To me, there is a matter that's more important than money, and that is peace of mind. If you don't play the game according to IRS rules, they can, and will, hassle the hell out of you. If I'm audited, my accountant takes control of the situation and he goes to visit the IRS. He tells me to stay home and prefers that I let him do his job. I can't put a figure on what that is worth to me. It's little things like this that make the difference between having a business that's fun and profitable—or just a hassle.

How to Shop Around

Here's a list of ten items to go over with a prospective accountant to help you decide if he's the one for you:

1. Can you afford him? Get his rates on what he would have charged you for last year's tax return.
2. Is he aggressive? Does he have that lean and hungry look? This is one area where complacency just isn't acceptable.
3. Is his office a mess? If he's disorganized and it looks like a bomb exploded in his office, don't have anything to do with him. This mess on his desk is representative of the disorganization between his ears. Stay away!
4. If you get audited, will he carry the ball for you—or give you a hand off? If the IRS calls, will he wet his pants?
5. Is he financially successful? If he doesn't know how to manage his own money, then . . . .
6. What are his professional credentials? Is he a CPA? PA? Enrolled Agent?
7. How does he suggest that you handle your bookkeeping? If he wants to do your bookkeeping, then he's a bookkeeper. Say "thank you" and find an accountant.
8. Get three references and check them out.
9. If you need him for something special, like incorporating, then ask him how many small businesses he's helped incorporate.
10. Does he deal primarily with small business? Stay away from the accountant who does a little of everything for anyone.

Do You Want Him on Your Team?

When I was a kid playing baseball, there was a fellow named Dave who was fantastic at bat and a great pitcher. But when we chose sides, I just didn't want him on my team. Why? He wasn't any fun, and he was very critical of others. I use the same criterion in choosing my accountant or anyone I hire. If he's going to be on my team, then we should both win and have a good time doing it.

RETIREMENT PLANS

The Keogh Plan

If you are a consultant operating as a sole proprietor or in a partnership, you may participate in a self-employed retirement plan which has been established under the Self-employed Individuals Retirement Act. This plan is often referred to as the Keogh Plan or HR-10, and allows you to accumulate your earned income and place it in a special retirement account which is operated on a tax-free basis. The Keogh Plan permits any partnership or proprietorship to contribute to the retirement fund up to 15% of one's earned income up to a maximum of $7500. The entire contribution to the retirement plan is tax deductible, and any earnings that you make through interest from these contributions are also tax free. The idea is simply that you pay taxes on the funds when they are paid back to you upon retirement when you will presumably be in a lower income tax bracket.

Work with Your Banker

In order to set up this type of retirement fund, you can work with your accountant or banker. There are even retirement plan specialists that can help you establish a plan that is just right for your business. However, if you are just starting a consulting practice and you are interested in this kind of plan, I suggest you work with your banker. He will be happy to give you all the information you need, because he wants your business and your retirement funds in his portfolio. He will do everything he can to help you establish your plan correctly.

Include Your Employees

When you choose the Keogh Plan, you must include all full tme employees who have three or more years of service with your firm. A full time employee is defined as a person who works more than five months a year and puts in more than 20 hours of work per week. The contributions that you make to the plan for your employees do not have to be the same dollar amounts as you put in for yourself. However, the *percentage* of the contributions have to be the same for yourself as the employees. Therefore, if you put in 5% of your earned income for yourself, you must also put in 5% of your employees' earned income into the plan. If you have a partnership, your partners are not obligated to participate in the program.

If you have employees that become part of this plan, you must remember that the money that you put into the plan in their name is their money, and you must pay it to them at some future specified date. This even applies to employees who leave your firm. If anything happens to an employee, he or his beneficiary receives the money.

Use the Profit Sharing Plan

As the president of a consulting firm making contributions to the Keogh Plan, there are two different contribution plans from which to choose: the profit sharing plan and the pension plan.

With a profit sharing plan, if you don't make any profits on a given year, you do not have to make any contributions. However, in a regular employee pension plan you are obligated to make contributions every year whether or not you show a profit. As a beginning consultant, I suggest that you only be concerned with the profit sharing type plan so you are not fixed with a large pension bill at the end of the year in situations where you do not make a profit.

There Is an Upper Limit

$7500 is the upper limit to the amount that you may contribute to your retirement plan. For example, if you are a consultant and you have earned $60,000 a year, you may contribute only $7500 and not $9,000, which is 15% of the $60,000.

There is one exception to the 15% contribution rule which I hope you don't have to utilize. If you earn less than $15,000 from all your sources, then you can contribute $750 or 100% of your self employed income, whichever is less. Of course, I suggest that you plan on making more than $15,000 a year, and thus you won't have to use this clause.

You are allowed to put more money into this plan if you would like to make a retirement fund larger. However, the extra money that you put into the plan above your regular contributions is not tax deductible. For example, let's say you make $80,000 a year and you contribute $7500 into the plan. This $7500 is tax deductible. You may also place up to $2500 into the plan and this $2500 is not tax deductible.

An important point with regard to the Keogh Plan is that you don't have to make any contributions during the year or even at year's end. You may wait until April 15th of the following year to make the entire contribution.

Where to Place Your Fund and How to Use It

You will not be able to use this money to buy or sell property. In particular, you won't be able to make loans with it or use it directly in your business. With this in mind, you should exercise your right to decide among various types of funds with considerable care.

In the creation of your fund you have a choice of the following: (1) United States Treasury Retirement Bonds, (2) A bank account which is set up solely for investment, (3) Annuities or life insurance purchased directly from an insurance company, (4) a trust account where the bank is the trustee. In this situation you are allowed to direct the trustee in the investment of the funds. There are a few other methods using certificates that you may purchase in setting up your fund, and you can get this information from your banker or insurance agent.

You are allowed to use the retirement fund as an income source after you have reached the age of 59-1/2, and you may not use the benefits beyond the age of 70-1/2.

You can take the money out of the plan as needed and it will be taxed as ordinary income. Hopefully you'll be in a lower tax bracket when you're doing this. Also, if you made voluntary contributions over and above the tax deductible contributions, you may withdraw these contributions tax-free.

There are some situations, however, where you can receive the benefits earlier than the age of 59-1/2. If you or one of your employees becomes disabled, then you may reap some of the benefits of the plan. If you or one of your employees dies after retirement and there is some money in the fund remaining for that individual, then the money must be distributed to the person's beneficiary within a five-year period. Note that if an employee leaves your firm for reasons other than retirement or disability, any accumulated funds that are set aside for him that are for his benefit must be paid to him at the time of his severance with your firm.

Withdrawal Penalty

If you have determined that you really need the money you may withdraw it, but there is a 10% penalty for premature distribution of the funds. (However, if you have placed money into the plan that is not tax deductible, that is, if you have placed money in the plan that is over and above the amount that you use for tax deductible contributions, then you may withdraw this money from the plan without any penalty.) You should not confuse this 10% penalty with a tax. If you withdraw the money it will be taxed at a regular tax rate, in addition to the 10% penalty.

Earned and Deferred Income

It should be noted that the income referred to for the Keogh Plan is *earned income*. Earned income is derived from personal services rendered as distinct from other kinds of income. For retirement plan purposes, earned income is the net profit obtained from your services. It is also referred to as "realized revenue" or just plain "earnings."

Earned income should be distinguished from deferred income. Deferred income is revenue received or recorded before it is earned. Examples of deferred income would be rent received in advance, unearned subscriptions, or money received by a consultant in advance as a special retainer, but not yet earned by the consultant. Some retainer type advances are considered to be deferred income if the income was put into an account with the understanding that it is to be drawn by the consultant provided he has worked for it. (Some retainers can be considered earned income if it is agreed that it is money that is placed aside just so the consultant could be called upon in case his client should need him, in which case this income would be considered earned whether or not he consulted to his client.) When you have a situation of obtaining deferred income, you may not use this income in the calculation of the deductions that are to be used for your retirement plans. Only earned income is allowed in this situation. If you receive retainers, you should review them with your accountant to determine if they fall into the category of earned income.

Pension and Retirement Plan Pros and Cons

As I said, you should be wary of setting up a pension plan that commits some of your working capital every year even if you don't show a profit. Pension plans are better suited for companies that have more income predictability than that which occurs in a consulting business.

Another fact you should consider is that as a consultant you will most

RETIREMENT PLANS

likely consult while in your sixties, and you will most likely still be in a high tax bracket when you take your money out of the Keogh fund. If you are just starting out in consulting, it might be better to reinvest your earnings into your own business expansion. Investing in yourself is one of the best investments you can make.

Annuities

An annuity is a retirement plan often obtainable from banks and insurance companies in which you make payments for a given number of years and then, after a certain determined date (usually at retirement) you receive income for a definite period or for the rest of your life. A straight annuity agreement is set up such that annual payments are made until a maturity date. After the maturity date, the annuity holding company pays you a regular income until your demise. Note that in some situations there is no guarantee of a minimum total payment. Also, the annuity has no cash value, and all payments will cease after you die. Due to low interest yields, annuities have become unpopular lately, and unless interest rates change, they will probably remain so.

Joint Annuity

Two people can take out an annuity. An agreement of this type is referred to as a "joint annuity," or a "joint survivorship annuity." It usually involves a husband and wife, and after maturity will continue to deliver an income as long as one of the parties is alive.

Deferred Annuity

A deferred annuity is set up such that payments to you from the annuity holding company begin after a pre-determined number of years. Note that this payment will start arriving even though the premiums themselves are still required and maturity has not been reached. In the deferred annuity, the company does not pay anyone if you die before the payments begin. For somebody going into a consulting business, this type of pension plan might not be advisable because of the risk.

Refund Annuity

In a refund annuity, the program is designed so that you or your beneficiaries will receive what you have contributed or more. If you die before this total amount has been paid back to you, then your beneficiaries will receive the rest.

Variable Annuity

A variable annuity is indeed what its name suggests—a variable. The annuity holding company makes investments such as stocks, the value of which will vary over the years. The annuity fund is designed to take any price changes into consideration. You never know until the maturity date what the exact value of the variable annuity is. When the stock prices have risen, you will receive a larger amount for your annuity, and if the stock prices decline, your annuity will be worth a lesser amount in due proportion.

TAX SAVINGS FOR CONSULTANTS

Self-Employment Tax

If the IRS considers that you are self employed, you must pay a self-employment tax. However, just because you call yourself a consultant, the IRS might not consider you a consultant if your work is primarily for one firm and they exercise a considerable amount of control over you. The fact that you bill an employer as an independent does not necessarily mean that you are independent in the eyes of the IRS. However, if you're consulting to a number of different companies there will certainly be no problem; there will be little question that you are an independent consultant.

Your self-employment income is the net profit from your consulting business and any other profits that you make from other ventures. These other business ventures do not include income from real estate or items such as dividends and interest from bank accounts or stocks and bonds. Also, if you sell property and make a capital gain on the sale, this is not considered to be a self-employment income. In 1981 the self employment tax was 9.3%, up to an income of $29,700. Thus, the maximum you would have paid in 1981 would have been $2,594.70.

If this is your first year of operation and you worked part-time consulting and full-time in your regular position, then you do not pay the self-employment tax. Note that you already had money taken out of your check for withholding and social security. If you were working full time and you earned less than $29,700, then you should pay the IRS taxes on the difference between your income from your job and the $29,700.

If you spent most of the year setting up your business and your income was under $400, then (if you had no other income) you're not obligated to pay the self-employment tax.

Are You Operating a Business?

Before one may make business-related tax deductions, one must meet certain IRS specifications as a business. Your consulting practice will be considered a business if it is an activity to which you devote a substantial amount of time and labor for the purpose of making a profit and livelihood. It would seem that when you operate as an employee, your activities would satisfy this definition for a business. However, the definition of a business has an impliction of independence attached to it. An employee/employer relationship seems to minimize the employee's sovereignty such that the IRS would not consider an employee's activities as a business. You can see from this definition of a business that if you pursue a hobby, it would not be considered a legitimate business. Combine that hobby with a demonstrable sovereignty and quest for a livelihood, and you may be able to convince the IRS that your activities, although they they appear as a hobby, are in reality a business.

Also, if you operate an activity which you call a business, but it does not realize a profit one out of every three, or two out of every five years, it could be classified as a non-business entity. Resulting losses that could be obtained from such an activity would not be allowable as a legitimate deduction. Note also that there are no statutes or rules which the IRS go by which could prevent you from operating more than one business.

IRS Definition of Independent Contractor

The following quote is taken from the IRS Form SS-8 entitled Information For Use In Determining Whether A Worker Is An Employee For Purposes Of Federal Employment Taxes And Income Tax Withholding:

> "The determination of whether a worker is an employee, for purposes of the Federal Insurance Contributions Act, the Federal Unemployment Tax Act, and the income tax withholding provisions of the Internal Revenue Code, is based on the usual common law rules applicable in determining the employer–employee relationship.
>
> "Under the usual common law rules, the relationship of employer and employee generally exists when the person for whom services are performed has the right to control and direct the individual who performs the services, not only as to the result to be accomplished by the work but also as to the details and means by which that result is accomplished. That is, an employee is subject to the will and control of the employer not only as to what shall be done but how it shall be done. It is not necessary that the employer actually direct or control the manner in which the services are performed; it is sufficient if he or she has the right to do so. The right to discharge is also an important factor indicating that the person possessing that right is an employer. Other factors characteristic of an employer, but not necessarily present in every case, are the furnishing of tools and a place to work.
>
> "In general, if an individual is subject to the control or direction of another person merely as to the result to be accomplished by the work and not as to the means and methods for accomplishing the result, the individual is an independent contractor. Also, if it is possible for the individual to sustain a loss, the individual is usually an independent contractor. An individual performing services as an independent contractor is not, as to such services, an employee under the usual common law rules. Individuals such as physicians, lawyers, dentists, veterinarians, construction contractors, public stenographers, auctioneers, etc., engaged in the pursuit of an independent trade, business, or profession, in which they offer their services to the public, are independent contractors and not employees.
>
> "Whether the relationship of employer and employee exists under the usual common law rules will be determined after the examination of the particular facts of each case.
>
> "If the relationship of employer and employee exists, the designation or description of the relationship by the parties as anything other than that of employer and employee is immaterial. If such relationship exists, it is of no consequence that the employee is designated as a partner, coadventurer, agent, independent contractor, or the like.
>
> "All classes, or grades of employees are included within the relationship of employer and employee. Thus, superintendents, managers, and other supervisory personnel are employees."

How Do You Decide What to Deduct?

There are many expenses and types of activities which will arise in the course of your business that seem to be difficult to classify. Is the activity a business activity or is it pleasure? After all, there is an overlap between these two activities. For example, if you combine a business trip with a vacation, it might not always be clear where the vacation begins and where the business activities end. This difficulty of separation has created much litigation and resulting tax literature.

Also, there are activities such as country club dues which, for many people, are primarily business activities and for others are pleasure. The legitimacy of the deduction for the same activity could vary from person to person depending on how his business objectives related to the activity. Therefore, it should be clear to see that the IRS has a difficult time in making absolute distinctions. Putting various activities into certain compartments that apply to all people does not work. Another example would be telephone calls to business associates who are close friends. Since the telephone call can serve many purposes, it's difficult for the IRS to make hard and fast rules in this area.

When you are considering the deduction of a particular expense, you should determine if the activity satisfies the following criteria:

1. The expense or activity should be an expense incurred in a business which is carried on by the person paying the taxes. Simply stated, the expense must be directly related to your consulting business.
2. Just because it's an expense related to your business does not mean that it is deductible. The expense is not allowed if it is a capital expenditure. Examples of capital expenditures which are not allowable are the cost for a building or particular equipment that you might use in your consulting practice.
3. The expense must be "ordinary" and "necessary." The word "ordinary" is a generality that must be interpreted from the way the courts have made the decisions. In many situations, the word "ordinary" is used to distinguish between capital expenditures and other types of expenses. The word "necessary" implies that the expense had to be incurred in order for the taxpayer to carry on his business successfully, i.e., make a profit. Again, this is a fairly general term.

The reader might find it interesting to see that the IRS can profit from using generalities and giving incomplete definitions. If a large portion of the public is afraid of being hassled by the IRS, they might not make deductions in places where they're not sure. Thus, of course, when a person is afraid of the hassles that can be generated by the IRS, there are many things that he won't deduct that might be legitimate deductions. One approach to this situation is to make a habit every year of studying a little bit more about taxes and eventually getting to the point where you understand what is a legitimate deduction and what is not so that you can legally minimize your taxes.

TAX SAVINGS FOR CONSULTANTS

Keep Records

Before going any further, I should emphasize one very important point. The key to taking full advantage of tax deductions lies in keeping track of all expenses and documenting each as thoroughly as possible.

After awhile, keeping track of such activities becomes second nature, and it really doesn't take much extra time on your part. You can have your secretary keep written records of all your expenses, or you can leave a dictating machine in the glove compartment of your car. As you spend money, you can note the data down and record what you spend your money on. You don't even have to have it transcribed until the end of the year, at which point you might find that you spent a considerable amount of money which would otherwise not have been accounted for.

Sometimes your employees don't keep sufficient records of their expenses, and such lack of records could lead to an eventual loss of money through an IRS disallowance. Therefore, have your employees keep a close record of their activities.

Along with having a record of everything for your daily activities, you should also have itemized receipts available for ready evidence. Sometimes a cancelled check is not sufficient, and an itemized bill along with the cancelled check is much better. For example, a cancelled check written out to a hotel is not nearly as convincing as a cancelled check in addition to a total itemized receipt.

Temporary vs. Indefinite Employment

Sometimes a consultant takes on a contract and then receives an offer from his client to become an employee. He should recognize that if he's away from home and deducting living expenses due to the temporary nature of the job, he stands to lose quite a bit by not having this deduction. As soon as you accept the job, you are no longer on a temporary basis, but fall under the category of being employed indefinitely. If the anticipated employment is a year or less, then you can consider the position temporary, but such a transition should be considered with caution, since losing a deduction as large as "away-from-home living expenses" can be a matter of many thousands of dollars.

One rare but still possible situation in consulting could occur if you don't have a legitimate home residence and you're just traveling from job to job on long consulting assignments. Note that you could be considered an itinerant worker in such a situation and might have all your living expenses disallowed. In such circumstances, you should make sure that you've established a legitimate residence so that your expenses will be truly travel and living expenses.

If you don't have a separate business office, then your regular home residence will be regarded as your center for business activity.

Example: You decide to take a consulting position which is a considerable distance from your home and, because it looks as if it's going to be long job, you decide to take your wife with you. You think that the job will take approximately one year, but you're really not quite sure, and it could lead to a position of indefinite length. You take an apartment in the area and try to deduct your living expenses for your first year.

Due to the fact that you have not made arrangements with your client with regard to the definiteness of duration of the position, the IRS could claim that you could go on claiming your living deductions and continue on your consulting position for an indefinite period of time. In an example such as this, the IRS most likely would disallow your living expense deductions.

Professional Deductions

Here are some examples of deductions which you can make as a consultant in private practice. This list is not complete, but it should give you a fair idea of the spectrum of items to deduct.

1. Office telephone
2. Office utilities
3. Office rent
4. Office supplies
5. Professional society dues
6. Supplies used in consulting
7. Supplies needed to finish projects that were not reimbursed by clients
8. Automobile expenses which are used directly related to the consulting and were not reimbursed
9. Office repairs and improvements
10. Liability insurance
11. Workmen's compensation insurance
12. Business-related accident and theft insurance
13. Wages and salaries paid out to employees
14. Fees to other consultants
15. Fees paid for secretarial work
16. Fees to anyone who helped you obtain business or clients
17. Interest on debts which are business related
18. Taxes on property that is being used for your consulting business
19. Equipment rental which is related to the consulting business
20. Computer time
21. Business related travel expenses

Library and Equipment

Your books and materials for a professional library are depreciable. The rate of depreciation varies from profession to profession. You should determine how long the serious items in your library are useful and then deduct on that basis. If you look closely at what you've bought over the past few years, you might find a considerable amount of the material is no longer applicable to your work. If this is the case, you might be able to deduct the depreciation of some of these materials.

Any equipment, instruments or books which have a useful life of a year or less are totally deductible. You don't have to put them into the depreciation category. These items can add up considerably, and you should have receipts for all of them. Remember, in order to be an immediate deduction, their *useful* life has to be a year or less. If you must buy books and equipment for the sole purpose of completing a contract, then you can deduct their expense.

A Consulting Office in Your Home

The IRS has become more strict over the last few years on deducting expenses that are incurred from your office at home. They have been particularly adamant that your home office be used exclusively for business activities, and that it be separate from the rest of your home. As your principal place of business, it can also be deducted if it is the place where you meet with your clients and carry on your normal consulting work.

Since technical consulting is a professional activity, this is an area where the IRS recognizes that a person can have a legitimate office in the home. If you are a professional engineer, indicate your P.E. number on your tax return, and state that your office is used exclusively for business purposes. If you are also using an office in another location, it can be questioned by the IRS, and you should work out the details with your accountant.

You should determine what percentage of your home or apartment is being used by your business activities (exclusively). Then, once you have that percentage, you can add all the items such as rent, utilities, heat, light, insurance, etc., and take the appropriate percentage for your deduction. You may also deduct any depreciation for furniture and equipment which you have. Also include a deduction for the depreciation of the section of your home which is being used as an office. Any money spent on setting up your office or making any repairs is, of course, a business expense and may be included in your deductions.

Traveling to and from the Office

A general rule is that you can deduct expenses from travel which occur in going from one business location to another business location. If there is an office in your home, then with regard to traveling, it should be regarded as a place of business. Thus, if you travel from your office (which is in your home) to another place of business, you can deduct that traveling as a legitimate expense. If, however, your home is not your principal place of work, then any cost traveling from it to a business location or from a business location back to your home cannot be considered a legitimate deduction. These expenses are referred to as "transportation expenses," and should not be confused with "travel expenses," since the IRS uses the words "travel expenses" to mean expenses incurred from an overnight trip.

For a technical consultant who does a considerable amount of driving to see clients, this could be a major tax saving area. It could become quite a hassle to have to drive to your office before going off to see a client. This is one of the areas where a considerable savings could be obtained if you can legitimately claim your principal place of business in your home or apartment.

If the following travel items are primarily business expenses, they may be deducted:

Meals
Hotel and Motel Expenses
Airplane Tickets
Railroad Tickets
Convention Costs
Cleaning Expenses (when traveling out of town)
Insurance Cost on Packages during Travel
Travel Expenses to and from Airport
Telephone & Telegraph Communications

Transporting Equipment

If you incur any extra expenses due to the transportation of equipment which is needed to get a particular job accomplished, you may also make deductions for these extra expenses.

Example: You have a private civil engineering consulting practice, and you occasionally have to transport a large amount of surveying equipment to a job site. You rent a trailer to transport your equipment and you then deduct the extra expenses incurred due to the trailer rental. The IRS has found this type of deduction to be allowable.

Travel and Entertainment Expenses—Business vs. Pleasure

When the IRS looks at many sections of your return, one of the questions which they must decide is whether your activities are primarily business or primarily pleasure. Expenses which you incur for personal reasons or for pleasure are not valid deductions. If you don't distinguish between business and pleasure, the IRS might have to do it for you, which will lead to hassles that you should avoid from the beginning.

At first glance it might seem odd that travel expenses can add up to significant amounts. Even entertainment expenses, which are deductible in certain circumstances, can be significant. First of all, there are a large number of business activities that you carry out with your automobile that add high mileage to your business odometer. Also there are a large number of meals or business dinner tabs which might be carelessly neglected merely because the consultant doesn't keep the receipt or doesn't pay for such items with charge cards which could assist in keeping a record.

Forty dollars spent per week on business meals or entertainment add up to a two thousand dollar deduction a year. When you're in the 40% to 50% bracket, these items certainly add up. Watch your consulting expenses carefully, and you'll probably be able to put an extra $500 in your pocket for travel, possibly another $500 in your pocket for entertainment, and save another few hundred for meals. It's not unreasonable at all to add an extra $1,000 to your net just by watching these activities closely. Until you sit down and look very carefully at the amounts you're spending on food, travel and entertainment, you might not realize the savings that are available.

Traveling for Business vs. Traveling for Pleasure

As I have mentioned, the IRS will allow you to deduct expenditures incurred during your consulting traveling. However, if these expenditures are made for a trip that was partly pleasure, then you're not allowed to deduct all the expenses.

Let's say that you take some time off from seeing your clients and you attend a convention to come up to speed on the latest developments in your field. You also decide to take a few days off and do some sightseeing. You're allowed to deduct the expenses which were directly related to the convention and your business activities. The expenses that are incurred through sightseeing and entertainment related to your pleasure trips are not deductible.

In another situation you might decide to do some vacationing, and during that vacation you do a little bit of business, or even attend a convention for a day or so. Can you make deductions when the original trip was primarily for pleasure? Generally speaking, the IRS will not allow you to make these business deductions on trips that are primarily for pleasure. You must be careful on this point, because a trip that was primarily made for pleasure can often be made to look like a business trip, and this is something that you want to avoid.

When Are Your Spouse's Expenses Deductible?

If your spouse accompanies you on your business trip, you must be able to show that any expenses due to his/her presence are related to a legitimate business purpose. The IRS will not allow expenses that are incurred if your spouse's accompaniment is made for pleasure or vacation, or if he/she only performs occasional services and is not critical to your business activities. Take, for example, the case in which your wife does some occasional typing for you on your trip. Occasional typing or occasional running of an errand is not sufficient for writing off the expenses for the entire trip. If, however, she is actively engaged in your work and acts as an employee, business associate or partner, and you can demonstrate that her presence is necessary, these expenses will be deductible.

If you can demonstrate that it is necessary for your business to have your spouse present for certain types of social engagements, and that his/her presence substantially contributes to the enhancement of your business, then the IRS might allow this type of expense. However, you should be able to demonstrate that this is the case instead of just saying so.

Conventions, Conferences and Seminars

This section only includes data related to conventions, conferences and seminars held within the United States. The rules and restrictions related to conferences held outside the country are more restrictive and will be handled in the next section.

If the conference, convention or seminar is directly related to the consulting practice or to some of the technical specialties in which you are involved, then there will be little question as to the applicability of these activities to your business and, consequently, the legitimacy of your deducting their costs. Note, however, that if the activities are held in a resort area, then the IRS might agree that the conference is directly related to your work, but they might want proof that you did attend the conference instead of just going to the area for a pleasure trip. If you attend a conference which is directly related to your business and part of this time is spent for pleasure, then you allocate the expenses for business and pleasure accordingly.

If there's any question on the allowability of the deductions, the courts will be concerned with the amount of time you spent with business activities compared to pleasure activities. They'll also be concerned with the location of the activities. Obviously, if the activities were not held in a resort area, the question of pleasure would not be questioned as strongly. The other factor that they'll be concerned with is how the convention is related to your business activities. There are many engineering conventions that are directly related to the type of work which you do, and in this situation you should have little trouble convincing the IRS that this was a business expense. You might find it helpful to take photocopies of your convention notes and place them into your tax records for future reference. Also, it would be beneficial to retain the convention program and indicate those activities in which you participated.

We are speaking strictly about deductions that occur in business-related conventions. Conventions that are classified as "fraternal" are not allowable expenses for deduction. Thus, if you belong to the Royal Order of the Moose, you're going to have to foot those convention expenses on your own. It won't matter to the IRS if you meet with a number of business colleagues or clients at your fraternal meetings; the fraternal activities will be regarded as non-business activities. Of course, if there are charitable contributions which you've made to such organizations, they will be an allowable deduction.

It is important to keep thorough records of your business expenses at conventions. Special attention should be paid to itemizing these expenses, leaving no room for doubt when it comes to adjudicating monies spent for business vs. pleasure.

Foreign Conventions

The IRS is more restrictive with how you handle your expenses on foreign conventions than for conventions held within the United States. You may only deduct the expenses for two conventions a year which are held outside the boundary of the United States, Puerto Rico, U.S. possessions and trust territories. If you attend more than two conventions, then you have to pick the two conventions that are most ideally suited for deduction purposes.

In addition, you must also provide the following information. If this information does not accompany your returns, you will not be allowed the deductions.

1. You must provide a signed statement indicating which days you traveled and the total number of days you attended the convention. The time they are interested in is your time while at the convention, and you do not have to include the time spent traveling.
2. A record of how you spent your time with regard to *scheduled* business activities. You should have this record for each day of the convention where deductions have been included.
3. The IRS will also want to see an agenda for the convention activities. This agenda should be, of course, related to the statement of your activities.
4. An officer of the convention's sponsoring organization should give you a written statement which verifies your schedule for each day's activities, and the number of hours which you attended the various seminar activities.

Transportation Expenses

Transportation expenses can only be fully deducted if more than half of the total trip (excluding travel days to and from the convention site) is devoted to business activities. If less than half of the total days of your convention time are related to business activities, then only a fraction of transportation expenses may be deducted.

Living Expenses

The restrictions under which you are allowed to deduct your living expenses are even more complex and stringent than the transportation expenses. The key to deducting living expenses for a particular day will depend on the number of hours a day that you have scheduled for the convention and your attendance at business activities. If you want to deduct the entire day, you should have records of at least six scheduled hours, or you should have a record that you've attended at least two thirds of the total hours of the scheduled program for that day. If only half a day's business activity is scheduled, then you may only deduct half of that day's expenses.

The IRS may put a maximum limit on how much you deduct for a particular day (even if you attended the convention activities for the entire day). For example, if you deducted expenses for meals, lodging and local travel which exceed the U.S. employee's *per diem* (for that locale), the IRS may decide that these excessed expenses will not be allowed. If you have any questions with regard to this, you should ask the convention's sponsoring organization. They'll have data on what the maximum government *per diem* rate for U.S. government employees is for that locale.

Sales-Related Expenses

If you act in the capacity of a salesman for your consulting business, or if you have someone who is a salesman, then you may make sales-related deductions. Examples of sales expenses include: expenses incurred in entertainment, meals purchased with customers, split commissions, general business activity expenses incurred in your office for the use of sales, travel that is directly related to sales, and overnight expenses that are incurred for sales presentations. Examples of non-deductible entertainment activities are: social and country clubs, restaurants and theatres, hunting, fishing, and sporting events. Any expenses included during these activities such as food, lodging and travel are usually non-deductible as well.

Some entertainment costs are deductible when they are related to sales, if you entertain customers just to get on their good side or obtain good will. But this alone might not be enough to allow the IRS to grant your entertainment expenses as being deductible. You must be able to show the IRS that through the entertainment you expected to obtain direct and specific business benefits. Preferably these benefits should be expected in the near future. Being hopeful for benefits to come in the distant future will be more difficult to justify. Further, for the expense to be deductible, you must be able to conduct business during the entertainment. Thus, if you're to take a client to a boxing match, it might be difficult to explain to the IRS how you conducted business during the match.

You must also be able to show that your conduct of business during the entertainment activity was your principal activity. Thus, going off to a ball game and having a few words over business is not sufficient. If, however, you decide to go on a weekend fishing trip with a potential client to discuss future aspects of business, and you're doing so in order to get away and have a day or two which is uninterrupted, you might be able to get the IRS to agree to such an activity. The whole point here is that you're using the entertainment atmosphere to promote the business instead of just going out and having fun and doing a little business on the side. For activities such as sporting events, theatres, night clubs, cocktail parties and athletic events, it is up to you to be able to prove that your activities with your clients are *directly related* to business. Otherwise, these activities and your associated expenses will be disallowed. Keeping strict, itemized records for such entertainment costs is crucial, and you should always include the specific business purpose of these outings in your documentation.

If you belong to a sports club, you must be able to demonstrate that your activities during the use of such a facility were directly business related. If you can't show the IRS that more than 50% of your club or association activities were for business, then they'll most likely disallow any attempted deduction for such activities. Many clubs have monthly fees in addition to the initial startup fees to join the club. In many cases the initial cost of joining such an association could be considered as a capital expenditure for your business, and you can deduct them at the end of your first year.

Automobile Expenses

Automobile expenses are deducted for traveling from one business point to another business point. If you maintain an office in your home and it is a legitimate office, you may be able to write off all your automobile business expenses, since your home (in this case, your office) is a business location.

There are two methods of getting a travel deduction. In the first, you may deduct 20 cents per mile for the first 15,000 business miles, and 11 cents for every business mile thereafter. If you have two or more vehicles that are being used primarily or wholly for your business activities, then the standard mileage rate should be applied to the total business mileage of all the vehicles in question. Toll and parking expenses are also allowed in this deduction. Keeping receipts for these items is essential.

The second method is to determine what the actual expenses are for operating your automobile for the year. The sum of expenses includes gas, oil, maintenance, depreciation, etc. Compare the two methods and use the method which will give you the largest deduction. In this second method, items which you can include as legitimate automobile expenses are:

Gas
Oil
Lubrication
Tires
Insurance
General Operating Supplies
Automobile Washing
Maintenance
Parking Fees
Batteries
Tubes
Interest on Car Loans
Automobile Taxes
Licenses
Motor Club Dues
Automobile Painting
Storage Fees

There are items which you cannot deduct if, when added to the automobile, prolong the car's useful life or value. These items fall under a capitalization category which means that their cost should be added to the value of the car and then, as you depreciate the car, you will be able to recover the value of the parts added as the car is depreciated. One would think that something such as tires would fall under this category, but apparently this is not how the IRS sees it. If you changed the top on your convertible, when it didn't need repair or you painted your car for the sole reason of having it look better, you might find that these would be treated as capitalization which had to be depreciated. Another example would be adding a tape deck to a car which didn't have one. If you go to repair a broken tape deck, this is definitely deductible, but putting a new tape into your car is a capital expense and it must be depreciated with the rest of the car.

If you're itemizing the expenses on your car, don't forget to include the taxes which you pay on your automobile and any interest on car loans. This can be a considerable expense. Note that all the registration fees are also deductible, provided that the registration fee is a property tax in your state. Of course, if you use the standard mileage rate deduction method, you are not allowed to include these items.

The expenses that are incurred from interest on automobile loans and automobile taxes can be deducted either under automobile expenses or under the non-business section of Schedule A, Form 1040. The part of the interest and taxes which are not claimed as a business expense may be claimed as itemized deductions on Schedule A. This does not apply to driver's licenses and automobile registration fees, unless those items qualify as personal property taxes in the taxpayer's state. If you decide to use the standard mileage rate deduction method, then you can put your total auto tax and interest on the itemized section of Schedule A.

Example 1

The interest on a car loan for this year was $1,000. Since the business use of the car was 75% of the total, $750 can be applied to the deductible expenses of the car. $250 could also be applied as an itemized deduction on the non-business part of the Schedule A, Form 1040. Note that in the case of the standard mileage rate, the entire $1,000 interest payment can be itemized on the non-business interest section of Schedule A.

Example 2: Expense Methods

On the first day of the year a civil engineering consultant buys a new auto for his personal use and consulting practice. Over the course of the year he drives 28,000 miles. 21,000 miles were business related (75% business, 25% personal). His expenses and depreciation were as follows:

EXPENSES

Gas ($.10 per mile)	$2,100
Oil	100
Lubrication	100
Repairs	430
Tires	600
Replacements (windshield wipers, etc.)	130
Insurance	430
Motor Club Dues	100
Licenses	50
TOTAL	$4,040

DEDUCTIBLE AUTO EXPENSE (.75 × $4090) .... $3068.

STRAIGHT LINE DEPRECIATION (3 year)

Purchase Price	$8,000
Salvage Value	$1,000
Value	$7,000

Depreciation Potential (1/3 × value)	$2,333
Depreciation Value (3/4 × $2,333)	$1,750

AUTO DEDUCTION

Expenses	$3,068
Depreciation Value	$1,750
	$4,818

STANDARD MILEAGE RATE METHOD

First 15,000 miles (× $.20) =	$3,000
Remaining (1,000 × $.11/mile)	110
Standard Deduction	$3,110

Thus, we see that standard mileage rate method yields smaller results than the expense/depreciation method when the automobile is still being depreciated.

Equipment Leasing

Many small consulting firms must rent equipment to handle specific jobs that don't occur often, since certain types of equipment are too expensive to buy. Examples of this type of equipment leasing would be: specialty computers, word processors, oscilloscopes, etc.

Leasing equipment falls into two categories. The first category is referred to as a "financial lease," and the second category is referred to as an "operating lease." In a financial lease the lessee holds the lease for a large portion of time compared with the useful life of the equipment, and often pays more in total lease payments than the equipment was initially worth. An operating lease is a more common form of lease in which the leasing term is short with respect to the useful life of the equipment. This type of lease is often cancellable at any time, provided the leasee gives the lessor an advance notice of cancellation. A typical operating lease might be leasing a piece of electronic equipment for three months and then returning it to the lessor. The total cost of the lease might only add up to 10% of the total value of the equipment.

You'll have little trouble with the IRS when you write off the deductions for an operating type lease. However, if you're involved in a financial lease, then the IRS might be concerned with whether or not it is a true lease, and may claim that you have entered into what they refer to as a "conditional sale." If you are planning to lease any equipment that might fall under the "financial lease" category, then you should review the leasing agreement with your accountant and determine whether or not your lease is indeed a "true lease." You should not rely solely on the information the lessor gives you regarding a tax deduction. A useful reference for this type of problem is a booklet entitled "Tax Advantages of Equipment Leases" published by the Commerce Clearing House, Inc., 4025 W. Peterson Avenue, Chicago, Illinois 60646.

14

Should You Form a Sole Proprietorship, Partnership, or Corporation?

The choice of the legal form for your business is important. This chapter will help you make the proper decision that is aligned with your career and business objectives. The sole proprietorship, partnership and corporation all offer unique advantages which you must understand, but each have their disadvantages as well. The choice of the form of your business will also have an influence on your tax strategy. A small investment in time for business planning now will pay many dividends in the near future.

PART I: SHOULD YOU FORM A SOLE PROPRIETORSHIP?

Keeping It Simple

When starting a new consulting business, you should be concerned with keeping things as simple as possible; as you grow and learn you can take on new responsibilities. One way of starting out is to work as a sole proprietorship, i.e., a one-owner business organization. Most small businesses in the United States today are under the form of a sole proprietorship and are usually limited to the owner and a handful of employees. The proprietorship is ideally suited for ventures which do not require large capitalization and which can be operated by a small number of persons. Thus, it's not surprising that most small consulting organizations in the U.S. are sole proprietorships.

Freedom of Action

You have freedom in a sole proprietorship by virtue of being the owner of the business and running the show. In addition, the sole proprietorship also experiences an extra freedom due to the fact that federal, state and local regulations are few in number when compared to the regulations associated with a partnership or corporation.

By being the sole owner of the operation, you'll be responsible for setting the various policies that are necessary, handling the financial and legal affairs, hiring and firing, running the office, etc. The list goes on and on. This does not mean that you can't hire someone else to handle some of these responsibilities, but in the sole proprietorship the final responsibility for the business definitely ends with you. By assuming all these responsibilities, you gain complete control of your business. When decisions have to be made quickly, you do not have to consult with a partner or a corporation board. You can make the decisions instantly, and this gives you flexibility. You're also free to seek the advice of others without having to check with someone else first.

The Limited Life of a Sole Proprietorship

A sole proprietorship is a legal entity which is represented by the owner. The life of the sole proprietorship is thereby limited by the life and wellbeing of the owner. If you form a sole proprietorship and you die or become physically or mentally disabled, your business as an entity becomes legally void, and thus some legal problems can potentially arise.

For example, your heirs or relatives could have difficulties with regard to establishing ownership and claims to the properties that you've acquired through your business. To prevent things such as this, you should always have an updated will and legal documents which state the transfer of ownership in case you become physically or mentally disabled or in the case of your demise. If you were running a partnership or a corporation and you became disabled or died, the business entity could continue either through the actions of the remaining partners or by the officers and legal structure of your corporation.

You Set the Standards

Have you ever experienced working for an employer who set production standards which you were embarrassed to represent? It's not a very pleasant picture, is it? The same thing can happen when you're working in a partnership or a corporation. If you don't share similar professional standards and objectives with your partner, you can find yourself in a situation where your business associates are demanding that you deliver a lower grade product. When you're the boss, you have the advantage of being able to determine exactly what level of quality you want your service to meet.

In consulting, the quality of your service is extremely important. Other companies are hiring you and paying you a goodly rate in order to receive high quality information and service. Often you're being brought into a company to help them solve their problems, and you might even be regarded as an expert in a particular area. If you have business associates who want you to dilute the service quality, you could severely damage your company and your future. On the other hand, when the entire operation is run by one individual, it is easier to guarantee a high quality professional and ethical service. You create the needed policies for your company, and you have complete control over making sure that all your company obligations are met. When the company's reputation is in your name, it's easier to keep every aspect of the company working as efficiently and effectively as possible.

You Will be the Chief Cook and Bottle Washer

When you're lacking help in the beginning, there are many tasks which you will have to perform yourself. You're responsible for finance decisions, bookkeeping, paying bills, collecting debts, advertising, looking for new business, handling present contracts and many other duties. At first, because you have little familiarity with the many positions and responsibilities that must be handled, you may find it a bit overwhelming. But in actuality these tasks are all more or less straightforward, and you will find them easy once you get accustomed to them.

Don't Hog the Profits

In books that help people start their own business, it's sometimes mentioned that one advantage of a sole proprietorship is that you don't have to share the profits. That's true; you certainly can keep the profits all for yourself. However, if you want to succeed, you'd better learn quickly that by sharing your profits with those who help you, you're going to expand a lot faster. When you're in a partnership or a corporation, there are usually written rules and policies on how profits will be distributed. This is not the case in a sole proprietorship. You have the option of holding on to the profits for yourself, if you choose to.

Recognize that a major disadvantage in a sole proprietorship is that you don't have access to the information and energies of other professionals in the ready fashion that you might have in a partnership or a corporation. There are, however, many people who will help you succeed in your sole proprietorship, and you should share the profits with them. These could be other consultants or even someone who is just giving you sound advice. If you get the opportunity, share some of the profits with these people in the form of commissions or bonuses added onto their original fees. If they've given you information that has helped you make more money, share it with them. By sharing your extra profits with them, they're going to be much more interested in your future success, and they'll contribute their maximum when you deal with them in the future. Even the smallest bonuses are appreciated.

The Unlimited Liability of a Sole Proprietorship

Professional liability problems have been increasing over the past few years in some areas of consulting. If you make a serious mistake as a consultant, you could be held liable for the consequences of your errors. Examples of such problems are in design or construction where there are errors of estimation or omission and they result in a product that does not meet the required specification. In some cases the product appears to meet specification, but then breaks down at a later time, and the design professionals who were involved in the project can then be held liable for any losses that the client incurs.

In most circumstances that arise in consulting, mistakes that are found are often correctable, and situations are cleared up to the satisfaction of all the parties without going to court. However, in consulting which is related to the construction industry, there are other forms of professional liability that can be much more serious and more difficult to handle. For example, in structural engineering an inadequate design could lead to injuries to individuals during construction or after a structure has been built. In this case, an individual who was acting as a sole proprietor and who was not covered adequately by insurance could find his business success in serious jeopardy. One of the major problems associated with a consulting firm which acts as a sole proprietorship is that the risks are not shared with anyone except the insurance companies. Sometimes this is sufficient reason for people to form partnerships. However, if you're not adequately covered by insurance, or if you take the wrong job which eventually leads to a liability suit, you're often not covered in a partnership any better than you are covered in a sole proprietorship.

When acting as a sole proprietor, if you found yourself in an unhealthy financial situation, your creditors have the right to try to force you to pay your debts by attaching your personal property. These situations are referred to as "unlimited liability." Unlimited liability is the major disadvantage of a sole proprietorship, and if you're planning on entering an area of consulting where you could be involved in liability type suits, then you might want to consider forming a corporation which can give you some extra legal shelter.

Shotgun Suits

Over the past ten or twenty years there has been a rapid increase in what is referred to in the construction industry as "shotgun suits." In a shotgun suit, architects, engineers, consultants and any other parties who have taken part in a project can be sued together on the basis that they were all partially involved in any errors that were committed that resulted in eventual loss to a client. Thus, in some types of consulting, even if you performed your particular job satisfactorily, you could still be involved in rather serious suits because of mistakes made by others.

Review the Liabilities in Your Specialty

Prior to forming a consulting practice, you should determine what kind of liability problems are associated with your particular type of consulting. There are many areas of consulting in which professional liability problems are not common. Consultants in these areas do not have to worry about the protection associated with operating in a partnership or corporation. But if you are working in an area where professional liability suits are rampant, then you should seriously consider forming a partnership, in which the liabilities are shared, or a corporation, in which you can obtain some protection for your personal property.

Whether you're dealing with a sole proprietorship, a partnership, or a corporation, you should still be thoroughly familiar with all types of potential liability problems. Make sure that you have enough insurance to protect against professional liability situations. Even though premiums for such insurance might seem high, you must recognize that you're protecting your firm and yourself from many potential hazards.

What to Avoid

The greatest hazard is not carrying the wrong insurance, but being unfamiliar with the necessary information related to liability problems in your own area of expertise. *Though liability insurance is expensive, knowledge of how to operate correctly is inexpensive.* Take responsibility for finding the proper information with regard to liability problems and you should have little problem in these areas. For example, avoid working on programs where poor workmanship is prevalent. Also, don't choose clients who have a history of suing their contractors.

Flying Solo

Along with a greater freedom of action and control, you can often find yourself in a position of being the sole decision maker, and sometimes this can be a direct disadvantage. It's great to have the freedom of action to make your own decisions and be your own boss. However, in some situations you're not going to have all the proper information to make correct decisions. In this case, you can wind up in a pretty sorry state due to the fact that you had no one to turn to.

You should always have access to an excellent accountant and a competent lawyer. You should also have associates to whom you can turn when you have questions. Just because you're a sole proprietor of the business doesn't mean that you have to be alone when making decisions. Hire people to help you make decisions; whether they are consultants or other professionals such as accountants or lawyers, you should have these people available. The time to make them available is not when you want to make a particular decision, but well in advance.

Creating an Informal Association

If you have a number of individuals who are interested in working with you in your new venture, but you believe that it is premature to commit yourself to the formalities of a partnership or a corporation, then it might be beneficial to create an informal association. In such an arrangement, you can all operate as individual sole proprietorships. Each can share the same office and facilities and individually obtain and handle his own individual clients. If situations arise in which you wish to work together on particular projects, you can form a joint venture arrangement whereby you still remain as independent business entities, but form, in effect, a temporary partnership to complete a particular contract. This will give you an opportunity to see what it's like to work with these people and gain some familiarity with their working standards. If you find that you work well together as independents and that your temporary joint ventures are successful, you'll have a solid foundation in the event that you wish to form a partnership at a later date.

Often partnerships or corporations are formed because of the economic influences that occur when starting your own company. Some people believe that the only way to handle such situations economically is to join forces with other people. However, you might find that simply forming an informal association will save you enough money such that a number of you can start off on your own. You might also find other independent consultants who are already in practice and who will be willing to work with you on an informal basis. Speak with a number of individuals to see if an informal association could be formed; you might find that it is much easier than you think. Three or four independents who are sharing a common secretary and office space can save a considerable amount of money. You'll be able to handle projects that you would not otherwise be able to tackle in the early years of your operation.

State Laws

Some states frown on forming informal associations and might look upon such an association as a partnership whether a written agreement is made or not. Before forming an informal association, you should consult with your attorney and find out how your particular association will be regarded by the state in which you reside.

PART II: SHOULD YOU FORM A PARTNERSHIP?

What Is a Partnership?

A partnership is a contractual relationship which is created on a voluntary basis by two or more persons to carry out business activities with the objective of making a profit. The contractual relationship between the partners is based on a written, oral or an implied agreement. The objective is to combine their financial, technical and administrative resources to create a profit, as well as a working environment in which they can expand their individual capabilities.

In a consulting situation, the partners are usually active members, although, as we'll see, it is possible to form partnerships in which some of the partners are inactive but still have a financial interest in the firm.

Generally speaking, the courts treat partnerships as single and separate legal entities in which each partner has partial control and is partially responsible for the activities of the enterprise. When forming a partnership to establish a consulting firm, it is best to form a contractual agreement in writing which delineates the responsibilities, duties, obligations and rights of each member. Clearly spelling this out prevents any potential misunderstandings.

CLASSIFICATIONS OF PARTNERSHIPS

General Partnerships

A common type of partnership is called a **general partnership;** most consulting partnerships fall into this category. A general partnership is one in which the partners are actively engaged in the transactions of the business and are publicly available to represent the firm. Each partner has at least some voice in the management, and in many cases the partners share equally in control of the partnership. They are governed and regulated by the Uniform Partnership Act.

Advantages

A general partnership has many advantages. It is easy to form a general partnership since it is only a matter of agreement between the partners, and the cost of forming a partnership is very low. A general partnership is not a taxable entity, and it can legally operate in more than one state without being required to comply with many of the legal requirements in each state. Another advantage of partnerships is that they are subject to less regulation than corporations (though they are subject to more regulation than a sole proprietorship).

One of the major advantages of a partnership over a sole proprietorship is the relative ease of raising capital and obtaining better credit. If the members of the partnership already have good individual credit, the banks will look favorably on the partnership agreement. The reason for this is that the individuals comprising the partnership usually have more combined assets that they can use as collateral. Also, if the business does not go well, the creditors can sue all the members of a partnership, whereas in a proprietorship the creditors only have claim against one person.

When it comes to raising capital, it is usually easier for a group of individuals to raise money than a single individual who's running a proprietorship. When people are trying to raise money to get a venture going, more faith is invested in a partnership than in the sole proprietorship.

Disadvantages

Some of the disadvantages of a partnership are that only a limited number of individuals may be considered owners. If one is trying to raise capital for the enterprise, then one might have to create a corporation and create stockholders in order to raise larger sums. Another disadvantage is that a partnership is dissolved when one of the partners withdraws from the group. This is in contrast to the corporation, which will endure even if the founder himself leaves.

Some people consider it a disadvantage to have each partner be liable for any business losses or problems. This again is in contrast to the minimum liability experienced by a shareholder in a corporation. In a partnership the partners are subject to experience unlimited liability with regard to contractual problems and civil suits.

In some cases, a problem with income tax can arise in partnerships. In a partnership a partner is taxed on his portion of the profits. He is taxed on his share of the profits whether these profits are distributed to him or not. In such a situation, a partner may have to pay income tax on money that was never received by him. This is very important to remember in starting up your consulting firm since this type of problem can arise when the profits are reinvested into the company for expansion. A consulting partner in this circumstance might need to have an independent means of income in order to pay taxes on income not received.

Partnerships Are Unpopular with Consultants

Over the years the partnership has become a less popular legal entity. For many consulting firms it has been more advantageous to either remain as a sole proprietorship or to form a corporation. A recent survey of approximately 5,000 consulting firms showed that less than 10% of the firms were partnerships, whereas the remaining 90% were split evenly between sole proprietorships and corporations. The fact that only 10% of the firms surveyed were partnerships should be a point of caution to anyone who is planning on forming a consulting partnership. In some cases it may be more difficult to start out as a sole proprietor or corporation, but it is often the more prudent route.

Limited Partnerships

A **limited partnership** is a legal entity which has many of the advantages of a corporation and some of the advantages of a partnership. It is formed by two or more persons and is created through a written agreement. In a limited partnership, the limited partner is only an investor; this type of partnership is often used to raise capital. The limited partner's rights are limited by virtue of the fact that *he cannot participate in the company's management.* (If he does, his legal status can change to that of a general partner.) The limited partner does have the right to inspect the technical and financial records of the business so that he can make judgments with regard to the future of the firm.

One of the advantages of the limited partnership is that the investor has limited liabilities. This is in contrast to a general partnership where the general partners share all or the majority of the business liability. Therefore, the limited partner has not only limited liability, but also limited control of the partnership's venture.

Another benefit to the limited partnership is that it allows the maximum use of certain tax advantages. In particular, accelerated depreciation can be used since a partner is allowed to deduct losses which are shown on the partnership tax return. Often, accelerated depreciation results in losses in the beginning years of the partnership. This loss is usually deducted instead of being carried forward. When the limited partnership shows a taxable gain, the partnership often changes its legal structure and is turned into a corporation.

Joint Venture

A **joint venture** is a business enterprise that is created when two or more persons agree to share their energy and expertise to make a profit in a particular business venture. A joint venture differs from a partnership in that a joint venture is project oriented. In a joint venture, the association formed usually involves a single business transaction or a very limited form of business activity. In a partnership, it's usually assumed that the partnership will continue on with the objective of making continual profits in many business transactions. With the joint venture, it is assumed that the lifetime of the association is finite and will end when the single transaction is over.

FORMATION AND DOCUMENTATION OF PARTNERSHIPS

Forming a Limited Partnership

The limited partnership is a hybrid between a partnership and a corporation, and it should not be surprising that there are a large number of legal restrictions with regard to the formation of such a legal entity. These restrictions fall under the Uniform Limited Partnership Act. In order to create a limited partnership, the parties must agree to and sign a document which contains the following:

1. The legal name of the partnership and the type of business which will be carried out under this particular name.
2. The location of the partnership.
3. The name and address of each partner with an identification of the general partners and the limited partners.
4. The predicted length of time that the partnership is expected to remain in existence.
5. A summary of the total investments made by each partner and any other additional investments from other sources. If any investments are to be made on particular dates in the future, this should also be recorded in the document.
6. The dates that any contributions or investments or profits are to be returned to the limited partner.
7. The percentage breakdown of profits that are to be received by the limited partner.

8. Any other rights that the limited partner might have, such as the right to add other limited partners or to substitute an assignee of his interests.
9. A specification of the relative rights of the various limited partners. If some limited partners have more or different rights to particular contributions or interests, then this should be stated.
10. The manner in which the limited partnership should be dissolved.
11. Any rights that the general partners have which are unique to the agreement.
12. A statement of how a limited partner could recoup some of his investment by means other than cash. For example, in some limited partnerships a limited partner might have the right to demand certain properties or other capital equipment rather than cash.

Regulations

The formal certificate must be recorded in the county where the limited partnership is to operate, and copies of this agreement are also to be filed in other communities where the partnership will be conducting business or has offices. Most states require that the limited partnership make a formal notice of its formation by placing a statement of formation in a local newspaper. When any changes are to be made in the document describing the limited partnership, then a new document must be written and filed. This is an extremely important rule to follow in a limited partnership. If changes in the limited partnership are made and a new document is not written and filed, then the formal limited partnership could cease to exist and the courts could then regard the partnership as a general partnership. There are some states that require that a limited partnership have the word "limited" associated with their name.

Working with an Attorney

If you want to form a partnership, you should first do some homework and come up to speed on some law with regard to the formation, operation and dissolution of partnerships. The objective should be to know enough about the law such that you can converse intelligently with an attorney; don't go to an attorney and say you want to form a partnership without even knowing what a partnership is. Go to the library and pick up some books on the principles of business law. Read the chapters on forming, operating and dissolving partnerships and get to know the language that's being employed.

Don't go under the misunderstanding that just because an attorney went to law school that he knows a great deal about partnership agreements. Don't take on an attorney who has not worked in situations such as yours. Don't be afraid to ask your attorney if he's experienced in these matters. Ask him to show you examples and copies of partnership agreements that he's already drawn up. If he doesn't want to show you these examples, you might want to consider going to another lawyer who is more open about his capabilities and experience.

Writing a Partnership Agreement

Legally, a general partnership can be formed without a written agreement between partners. However, I cannot advise going into any form of a partnership arrangement without a thoroughly written agreement which outlines each partner's rights, responsibilities and duties. Fortunately, the complexities of starting a consulting firm are not high, and you can usually write a successful partnership agreement which accurately identifies everyone's rights, responsibilities and duties.

One of the most important parts of a partnership is not the written agreement, but what the written agreement represents. You and your partners should have common business objectives that are well outlined and understood by everyone involved. Furthermore, you should be thoroughly familiar with the professional standards of each partner and how they relate to running the consulting firm. There'll be many ethical questions that arise in running the business, and unless you work with partners who share similar standards, you could find yourself in fairly serious trouble even before the business gets started.

In order to write an effective partnership agreement, every partner should have a thorough understanding of how the business is to be run. Partnerships can get into trouble by not really knowing the details of how the partnership is to be run. If you're starting a consulting firm for the first time and you really don't know the details of its operation, you should consider forming a partnership agreement that is valid for a fixed period of time, say six months or a year. Agree that at the end of that time, when you know a little bit more about the business, that the agreement can be rewritten and a new agreement will emerge with the rights, responsibilities and duties of each partner identified in greater detail.

The partnership agreement should include the date of the agreement, names and addresses of the partners, the nature of the business, the formal name of the partnership, its address and the addresses of any other offices.

You should get together with your future partner and determine what your specific goals are for forming and operating the business. Your specific business objectives can then be identified in greater detail. If you can't identify your objectives with regard to finance, law, management, technical and sales type activities, then it could be that you haven't worked out your business plans in enough detail. You're not going to be able to operate a successful partnership if you don't agree on your business objectives.

Make the Agreement Functional

Your attorney can help you form a legally solid partnership agreement. However, the partnership agreement should also be functional. Just because you have a competent attorney does not mean that you will have a partnership agreement that is functional. Your attorney does not know the details of the operation of your business, and if these details are not included in the partnership agreement, then the partnership agreement will only be legally valid. Your objective is to create a document that has functional practicality. If you can't refer to and use this document in operating your business, then its value is decreased to the point that it will only gather dust.

You should spell out in as much detail as possible what role each partner is to play in operating the business. You should break the business up into various sections such as legal, finance, sales, technical, etc. State specifically who will be responsible for carrying out the duties of those respective sections of the company. Keep in mind that when someone takes on the responsibilities of a particular section of the company he should also have the necessary authority to carry out his responsibilities as needed.

Designate a Senior Partner

One of the partners should be designated as a senior partner. This partner should have the responsibility for making most of the major business and financial decisions for the company. If he has particular veto rights, this should be mentioned in the agreement. If you enter into a partnership agreement where major decisions have to be shared and made between two or more people, you might get yourself into trouble. Unless the partnership is formed around one strong and capable senior person, the decision making on important matters could lead to very wide and deep-rooted disagreements. If you are going into a partnership with someone, you should be going into the partnership because everyone has particular expertise that they can bring to the group. The expertise of one of the individuals should be strong leadership.

One of the reasons for the failure of partnerships is the inability for partners to come to agreement on who is going to lead the operation. If you can't come to an agreement on a leader, you'd better review your objectives again and find out if they are really practical. Note that in a sole proprietorship and a corporation one finds well-identified leadership, and that over 90% of all consulting firms fall under these two categories. The reasons for this are not hard to find. Companies that succeed are often run by strong individuals who are capable of leading the company in the direction of business prosperity.

Partner Investment

The agreement should indicate how much each individual will contribute to the partnership. The contributions can be in the form of cash, property, technical equipment or labor. There are no laws that require a partner to make equal contributions. If contributions are to be made to the partnership over a period of time, then the contribution schedule should be written in detail.

If you have only one partner and both of you are putting in equal amounts of time, energy, money and equipment, you might come to a rather easy conclusion that the profits are to be shared on a 50–50 basis. However, in many situations there are more than two partners, and the percentage of ownership of the individuals varies. You might have one person putting up a considerable amount of money to start the business, and another individual with a great deal of technical expertise investing very little money. A written agreement must be made to show the relative value of each type of contribution. For example, there are some situations in which an individual has a technical expertise that is in such high demand that he might obtain 75% control of the partnership and still invest nothing in the formation of the company. No situation that is agreed upon should be considered strange if the relative value of the contributions are written down and agreed to. The important point here is that you should spend some time identifying what contributions are made by each individual and what form contributions take.

In any case, the time will come in the partnership where profits or losses are to be distributed to the various members. Needless to say, this aspect of your partnership agreement is very important, and you are, of course, legally bound to the agreements you have made.

Salary and Drawing Accounts

There are no laws that require that partners make equal or simultaneous withdrawals of funds from the company. The procedures for any withdrawal of funds should be determined beforehand and written into your agreement. Unless you anticipate what types of withdrawals are going to be required and have a firm understanding in such matters, you could easily find yourself in rather serious misunderstandings very early in the development of your consulting company. If you have a withdrawal agreement properly written, you will be able to prevent situations in which one of the partners can withdraw substantial funds that could hinder the operation of the business.

Partnership Termination

If one of the partners dies or becomes physically disabled, or simply wants out, a section should be written stating how these various situations should be handled. If there's a possibility of one partner's buying out the other partner, the facts and figures and particular amounts should be indicated. When a partnership breaks up, it is said to be "dissolved" or go into "dissolution." A dissolution can occur when a new partner is admitted to the arrangement or when major changes are made in the original partnership document. A dissolution can also occur when one of the partners is expelled from the group, and also when there has been a violation of the partnership agreement. Just because an original partnership agreement stipulates the duration of the partnership, it is not true that the originally stated duration is the only legal duration possible.

There are many places in a partnership agreement where your attorney's advice is extremely important, and partnership dissolution is one of them. Your attorney should review with you and your potential partners all the various aspects regarding ending the partnership. Detailed clauses should be written to assist you in forming an effective termination.

What will happen if you form a partnership and after six months of business your partner gets an excellent offer from another company and wants to leave the partnership? What happens if you get sick and you can't perform your responsibilities? What happens if one of the partners in your operation dies or becomes disabled? The more detailed the section on partnership termination, the higher the probability will be that you'll be able to avoid any legal difficulties in cases where the partnership has to be dissolved. You should also specify in your termination section how various insurance matters are to be handled. Sometimes it's helpful for a partnership to carry insurance which covers the life of each one of the partners.

The Firm Name

The law does not restrict a partnership on the form of the firm name, and generally, the partners can decide on any company name that they desire. In a consulting firm which has a small number of partners, it has become effective to use the last names of the partners involved. This is sometimes very beneficial for an advertising point of view in cases where the partners each have significant reputations and the joining of their last names is a message of confidence and reliability.

Partnership Responsibilities

Since a partnership, just as a sole proprietorship, is considered as a legal entity which is inseparable from the partners, the individual partners are responsible for the actions of the rest of the partners. Each partner is regarded as an agent for the partnership. He, therefore, has the power to bind the partnership into contracts by his actions. The Uniform Partnership Act provides for contractual liability in the case of a partner who carries on business in a usual fashion. If the partner acts or contracts in such a way that is detrimental to the partnership, the other partners can be liable.

A good rule of thumb is to remember that any actions you take can affect your other partners just as if they had performed the act themselves. You, as a partner, are responsible for any legal, technical, or financial obligations that are generated by one of the other business partners. Therefore, if your partner generates a loan for the business that you didn't agree to, you may be responsible for paying back the loan, even though you didn't sign the papers. Likewise, if your partner gets into any legal hassles that are outside your designated responsibilities, you will could be legally liable for his activities.

PART III: SHOULD YOU INCORPORATE?

What Is a Corporation?

A **corporation** is an artificial, intangible, legal entity that is created within and by the law. It is an entity which is created for and by a group of individuals who have decided to unite and transact business under a common name and for common business objectives. Since the corporation is an independent legal entity, it's existence is not affected by the death, incapacity or activities of the individual members. Like an individual, the corporation can sue or be sued. Also, the corporation can own various forms of property.

There are many states in whose laws the word "corporation" is not used; instead, the laws use the word "person." In situations in which it applies, the word "person" takes the place of the word "corporation," and the corporation can often be treated as a person in a court of law. This, of course, does not apply to all laws where it is clear that the word "person" refers to an individual and not a business entity. Although the corporation has many of the legal rights of a person, it is not considered a citizen in that the corporation does not have many of the privileges that are granted to citizens through the federal Constitution.

The corporation differs from the proprietorship or partnership in that it is created by statute law. This law can vary from state to state. In a proprietorship or partnership the business is represented by the individuals and is not a separate entity that is created by law. States in which the business is incorporated require that the corporation produce reports on its activities. The states can and do levy fees on the corporate operations.

Stockholders

The corporation is usually controlled by a group of stockholders who hold the majority of stock in the operation. The stockholders exercise their power through their voting rights. In many cases, many small and large stockholders of a company take little interest in the actual operation of the corporation when it is making money. When the company starts to do poorly, then a larger percentage of the stockholders usually come to the front and often ask for a change in the officers of the corporation.

The Powers of the Corporation

The corporation is somewhat like a person; it is an independent legal entity. Unlike an individual, it is limited in its powers. However, it does have certain "implied powers," and among these are:

1. Power to buy and sell real or personal property.
2. Power to sue or be sued under the name of the corporation.
3. Power to enter into agreements and make contractual commitments with individuals and other business entities.
4. Power to make by-laws for corporate activities.
5. Power to borrow or loan money for business activities of the corporation.
6. Power to appoint officers.
7. Power to expand and acquire other businesses or rights.
8. Power to retain undistributed profits and not to pay profits to the owners or shareholders.

Limited Liability

One of the primary advantages of a corporation has to do with "limited liability." In a corporate structure, each stockholder is only liable for the debts of the corporation to the extent of the stockholder's investment. If you, as the stockholder, invest $1,000 into a corporation and the corporation then goes bankrupt, the most you can lose is $1,000. Though a corporation can be sued, the individual stockholders are not liable if the corporation of which they are a part loses the suit.

Transfer of Ownership

Another advantage is the simple transfer of ownership. Anyone can buy and sell stock for a corporation at any time, provided the stock is available to be bought or sold.

Tax Advantages For Corporations

For the businessman whose yearly profits are in the $15,000 to $20,000 range, the issue of taxes is not of paramount importance. However, when a person's profits exceed $60,000 per year, his largest single expense is taxes. Often, corporations are formed for precisely this reason; there are significant tax advantages when profits are stably high and the person is certain he wants to continue the operation.

Incorporating creates a second taxable entity in addition to yourself. Starting on the day you incorporate, the corporation is taxed on the profits it receives. Consequently, strategically timing your incorporation can be very important. Your financial advisor can help you on this.

There are numerous other benefits to forming a corporation. As I have mentioned elsewhere, the upper limit for contributions to pension plans in 1981 was $7,500 per year for sole proprietorships. For corporations it was between $35,000 and $40,000. Whereas in a sole proprietorship, health and legal plans come from taxable income, in a corporation these plans are a tax deduction for the corporation, and are not taxable to the employees covered.

Though corporations are a bit of an administrative hassle, the tax savings can be more than mitigating.

Double Taxation

In a proprietorship or partnership, the individuals involved pay a Federal Income Tax on the profits. One disadvantage to a corporation is that it must pay a corporate tax, and the individual stockholders who receive dividends are also taxed. This situation is referred to as double taxation and can only be avoided by either paying no dividends to the stockholders or by forming a sub-chapter S corporation (covered below). In the sub-chapter S corporation, the corporation itself is not taxed, but the individual stockholders are taxed as if the corporation were a proprietorship or a partnership. Eliminating dividends is not a complete answer to double taxation. In this case the IRS can issue tax penalties when unpaid dividends occur.

Closed Corporations

A closed corporation is a private legal entity in which all or most of the stock is owned by a small number of persons. The stock is not available for sale to the public. Frequently, the owners of a closed corporation agree not to sell their stock without first offering their shares to the other stockholders in the corporation. In many consulting firms there is very little reason to go public and issue large amounts of stock. The closed corporation is easier to operate and has many of the advantages of the corporation without having to worry about a large number of stockholders to whom you must constantly report.

Another characteristic of a closed corporation is that the small number of stockholders often are active in the management and affairs of the corporation. In many cases the stockholders are also directors. Corporations that issue stock to the general public are often referred to as "open" corporations.

CREATING A CORPORATION

What Does It Cost to Incorporate?

Licenses for corporations are not obtained from the Federal Government. The right to incorporate is obtained from the state in which you wish to conduct your affairs. Though there is some degree of uniformity throughout the states, each state has its own unique laws regarding incorporation. The requirements of incorporation vary from state to state, and so do the incorporation fees. In some states the incorporation fees are as low as $50, and in some states the same fees can run into thousands of dollars. If you're planning on incorporating, you should first determine if you'll be doing business primarily in one state or in many states. After you've determined where you'll be operating the most, you should then contact your attorney and go over the pros and cons of filing within that state, or possibly within the state of Delaware.

If you incorporate in the state of Delaware and you go through all the legal paperwork yourself, you could probably do the entire job for under $150. If you decide to incorporate within your own state, it could cost you considerably more. For example, if you plan to incorporate in the state of California, it will cost you approximately $500 if you do all the work yourself. If you plan to retain an attorney, you could spend $500 to $1500 extra.

Methods of Forming a Corporation

If you want to incorporate your consulting firm to handle liability problems or to take advantage of some of the other benefits of corporate structure, there are many methods of going about it.

A first method is to remain fairly ignorant of the whole situation, retain an attorney, and have him do the entire job for you. What you will obtain from this method is precisely what you put into it—very little. After you've spent some money and gone through the formalities of dealing with your attorney, you'll have a corporation. But will you understand it and be able to use it to its maximum potential? No. If all you've done is worked with an attorney and let him lead you by the hand through every step of corporate formation, you'll probably be completely confused by corporate matters, and incorporating will be more trouble than it was worth.

A second method is to continue studying business and corporate law such that you can discuss the pros and cons of corporate formation with your attorney, and interact with him in an intelligent and positive way. In this case, if you've done your homework correctly, you'll be able to ask him the proper questions, and when he leads you through the steps of forming a corporation you'll be able to take full advantage. If you are to reap the most benefits from this technique, you'll have to take complete responsibility for your understanding and not let your attorney do all the work for you.

The third technique is to go through all the steps of forming a corporation yourself. There are "how-to-do-it" books on the market. An example of one of these books is: *How To Form Your Own Corporation Without A Lawyer For Under $50*, by Ted Nicholas (Enterprise Publishing Co., Inc., 1000 Oakfield Lane, Wilmington, Delaware 19810).

Using the "do-it-yourself" method is one of the most beneficial and effective means of learning the ins and outs of corporate law. There's no better way to learn than to go through the steps yourself and form your own corporation. In many cases, you'll be able to save up to $2,000. But, this is by no means the greatest benefit. Getting in there and getting your hands dirty is by far the best way to really find out what's going on with regard to corporate operation. It'll take you a little longer and you're going to have to take a little bit more responsibility, but if you've got the time, you should seriously consider this technique.

Don't worry about making minor mistakes in setting up your own corporation. The steps you must go through are very straightforward. Don't let anybody kid you. The details with regard to a corporation are more in number than a partnership or a proprietorship, but the corporation only will remain a mystery if you don't tackle the details and learn what it's all about.

One other good technique is to go through all the steps of incorporating yourself, and then before you send in the final papers, work with an attorney and have him check your work. With this method, you are guaranteed to learn the most, and you'll have the security that you've done your paperwork correctly. The financial cost for this method will be a little bit more than incorporating solely by yourself, but the extra attorney's expenses are certainly worth it. This method will also give you an opportunity to interact with an attorney in such a way that you're learning the ropes at the maximum rate.

The Subchapter S Corporation

When we reviewed the limited partnership, we saw that it was a hybrid between a corporation and a partnership with respect to the subject of liability. We saw that the limited partnership had the advantage that the most a limited partner could lose was the total amount of his investment. Because he was protected by a limited liability, he also had no share in the management, i.e., he had a limited control. A similar situation exists with a hybrid which is known as the Subchapter S corporation. The Subchapter S corporation is also referred to as a tax hybrid, because it is taxed in a manner that is similar to the taxation of partnerships. In matters other than taxation, the Subchapter S is treated similar to a traditional corporation.

The Subchapter S corporation has the same structure as the regular corporation and offers to the stockholders a limited liability protection. A Subchapter S corporation is considered a legal entity, but the corporation as such pays no income tax. Instead, like a partnership, the profits of a Subchapter S corporation "pass through" to the owners. The owners are then taxed at their ordinary individual rates.

Generally, the Subchapter S corporation has been found to be a very reliable method for any company that is operating on a loss, since the losses are immediately deductible on the individual returns of the shareholders. When profits are being made on a Subchapter S corporation, there are very stringent restrictions as to how these profits are to be divided among the shareholders.

Unfortunately, the laws surrounding the Subchapter S corporation are very involved, and space does not permit a treatment of them here. Many of the federal requirements on a Subchapter S corporation are outlined in an IRS publication No. 589 entitled "Tax Information on Subchapter S corporations." I will just say that for some consulting businesses the Subchapter S corporation is ideal and should be seriously considered.

Delaware Corporations

If you're considering forming a corporation for your consulting practice, you should examine the advantages of forming your corporation in the state of Delaware. The Delaware laws are very favorable to corporations. Many thousands of corporations in the United States are Delaware corporations. Many of them do not transact any business in the state of Delaware itself, but just use the state of Delaware as a founding state for their corporation. There are many reasons for incorporating in the state of Delaware. Some of these are as follows:

1. There's no income tax in Delaware for companies which are registered in the state of Delaware. Also, there are no inheritance taxes or taxes on shares held by non-residents.
2. Delaware is a state which has no securities law.
3. There is no minimum capitalization requirement for forming the corporation. Some states have capitalization requirements of at least $1,000. In the state of Delaware, a corporation can be organized with no capital at all.
4. One person is allowed to hold the offices of President, Treasurer and Secretary. This one person is also allowed to be all the directors. In some states it is required that at least three officers be present for the formation of the corporation. If you desire to form a

corporation based on your name alone, you can do so through a Delaware corporation. This is ideal for the consultant who is working alone.

5. A Delaware corporation can issue a no-par common and preferred stock. The price level on this stock can be fixed by the directors of the corporation.

6. Corporate meetings do not have to be held in the state of Delaware. All meetings and records can be held outside the state.

7. Since a very large number of corporations are established in the state of Delaware, the state has a very well-established and sophisticated set of laws that are relevant to the formation and operation of corporations. The court system is very knowledgeable about corporate activities, and if you have to go to court, you can rest assured that the legal proceedings will be carried out with less confusion than in many other states. The state of Delaware obtains a very large portion of its income from corporate-related circumstances, and the courts have been found to be fair in making many corporate-type decisions.

8. The state of Delaware does levy a franchise tax on corporations. A franchise tax is a fee charged for forming a corporation and gaining the right to operate and be established in a particular state.

9. The Delaware corporation may be formed entirely by mail. There is no requirement that you have to visit the state or conduct any business in the state. Corporate activities and formal corporate meetings can be held anywhere outside the state.

10. Shareholders are protected from the liability of corporate debts, and the officers and directors of the corporation are in some circumstances indemnified.

11. A corporation formed in the state of Delaware may be a "perpetual corporation," and may operate through stockholder voting agreements. They may also operate through voting trusts.

12. Formal meetings are not required. The directors of a Delaware corporation can alter the by-laws of the corporation and may act formally by unanimous written consent. This is important for small consulting firms that have neither the time for nor the interest in corporate meetings.

13. The Delaware corporation is allowed to hold and own securities of other corporations and other types of property. The holding of other securities can be carried on outside the state. The corporation is allowed to purchase its own stock, hold it, sell or transfer it.

14. The corporation may operate different types of businesses in many different combinations.

15. With the exception of taxes, the stockholder liability is limited to the amount of stock that is held in the corporation.

Piercing the Corporate Veil

There exists a notion that individuals who form a corporation are protected from the law and creditors. This is not true, and you should not be misled into believing that a set of incorporation documents are going to protect you in all situations.

Directors and officers are entrusted with the management of the corporation. The stockholders expect the management to act in such a fashion that the corporation will benefit. The first and foremost rules of action should be in favor of the corporation and its stockholders. All actions should be within the law.

For willful or negligent acts the directors or officers are held accountable. A corporation is a legal entity that can sue or be sued. The directors can also sue or be sued. The director is not only accountable to the public with whom he deals, but he is also accountable to his shareholders. In principle, your private holdings cannot be attached in a suit if you operate through a corporation. In practice, you might lose your shirt. Be careful and operate with the highest business principles, and don't try to use a corporate veil to cover up any misdoings.

Directors of corporations are not personally liable for corporate illegal acts just because they are directors. To be personally liable, it must be demonstrated that the directors were involved in the wrongdoings. Directors and officers are not liable for errors in judgment, provided it can be shown that they have acted in good faith and exercised reasonable prudence and skill.

DECIDING AMONG THE SOLE PROPRIETORSHIP, PARTNERSHIP, AND CORPORATION

As you remember, many of the aspects of a proprietorship and a partnership are similar. Most important with regard to consulting is the fact that in a proprietorship or partnership, the owners of the company hold maximum liability. This is a very important factor in forming the business. If you're in a type of consultantcy in which there is a large number of suits and your liability is high, then you should seriously look at forming a corporation. The facts that should be balanced between a proprietorship/partnership vs a corporation are:

1. Taxation
2. Liability
3. Control
4. Continuity of legal existence
5. Power to sue and be sued.

In reviewing the pros and cons and making a decision based on these factors, you should consult your attorney and accountant. However, when you must make decisions about forming a corporation due to liability problems, you cannot rely solely on your attorney's advice. The most important problems with regard to product liability and various suits have to do with the type of consulting you're going to be involved in. It's up to you to do a thorough investigation of the types of suits that you can expect. You should make use of insurance companies to determine what your liability problems are going to be.

Decide Who Will Run the Show

One of the most important aspects in deciding about the form of your organization will be related to who is going to control the business. Unless you determine the amount of control you want in the business in the beginning, you will not be able to make the proper decisions on the formation of the organization. If you or one of the potential associates desire a major control of the business, then you'll have to recognize that this one factor alone can determine whether or not you form a sole proprietorship, partnership or corporation. It also will be a major factor in how the corporate structure or partnership is drawn up. If you desire to be the controlling factor in your organization, then you should explain this to any potential business associate. You must have an exact understanding of what aspects of the business you want to control. If this is not agreed to in the beginning, you might choose the wrong form of an organization and will run into business problems related to control.

15

Start-Up Projects to Do in Your Spare Time

Now that you understand many of the basics of successfully running your own consulting practice, let's take a look at what you can do now to get going. It is what you do now, in your spare time, that will make the transition from salaried employment to consulting a smooth one. Working effectively in your spare time can make the difference between a smooth hassle-free first year and a catch-as-catch-can nightmare.

Suggestion 1: Decide What to Do About Your Present Job

One of the biggest mistakes you can make with regard to starting your own consulting business is to leave your present job prematurely. There's plenty to do in preparing for a new consulting career. Before you leave your job, you're going to want to collect all the information that's necessary, go through all the required tasks in setting up the business and then give your employer a healthy notice that you're going to leave so he'll have plenty of time to replace you.

The first thing you can do with regard to your job is to recognize the value of your present employer. Your employer has hired you in good faith to do an excellent job and to produce the necessary products of your position. Since he is paying you a fair salary, he expects and deserves the best from you. Use whatever relevant tips you've garnered from this book thus far and any other pertinent information to improve the productivity of the company. Under no circumstances should you spend time during the day preparing for your own consulting practice. Your present employer should not be paying you to go off and start your own practice. Working nights and weekends will be sufficient.

After investigating the possibilities of consulting and setting up a plan for going into business for yourself, you should have a pretty good idea of how long you'll have to remain in your present position. Work part-time for a few months in setting up your practice. Then, a few months before you're about to leave, contact your employer and explain to him that you're planning on going into business for yourself. Explain to him that you want to maintain a good relationship with him and the rest of the staff, and that you'll do everything in your power to make your transition an easy one. Give him as much notice as possible so he can find a competent replacement for you. It might even be beneficial to have your replacement work with you for a while so he can get to know the ropes.

If you make the transition properly, your present employer might offer you a consulting contract and he will remain a good client.

Suggestion 2: Decide on the Legal Structure of Your Business

Before you go into practice, you'll want to make the decision of whether you should have a company that is based on either a sole proprietorship, partnership or corporation. This decision can only be made correctly if you understand your objectives and the assets and liabilities of each type of organization. The chapter on this will provide a good start.

Don't commit yourself to a partnership or to forming a corporation with anyone until you have made a thorough investigation of all three types of organizations. In addition to these three, you'll also want to consider the possibility of temporary partnerships called joint ventures. These are often used in consulting and can give you the opportunity of working with someone on the basis of a partnership, and you will be able to see how you get along. If working in a joint venture works out, you might find that you have a solid foundation for working as a partnership or starting a corporation.

As I mentioned in Chapter Fourteen, you should make a thorough review of the various advantages and disadvantages of the three types of organizations. It should also be kept in mind that too often people form partnerships or start corporations with associates with whom they shouldn't be dealing. It doesn't matter how long you have known a particular individual. Knowing him as a friend or working with him as a fellow employee does not necessarily mean that he should be a business partner or should be a founding member of a corporation. You must know what your business objectives are. If forming a partnership or corporation satisfies many of your financial objectives, then you should go ahead and start that type of organization. But it might take a few months before you really understand what your specific business objectives are. *Do your homework and then start talking to your friends.*

Suggestion 3: Learn About the Contracts You'll be Using

As a consultant, you'll be entering into formal contractual agreements with your clients. These contractual agreements should be understood by you in every respect. You should first review your basics of contract law and then determine how these basics apply to consulting.

Get copies of contracts that apply to the type of consulting that you perform. There are professional societies in almost every discipline that have gone to great expense to establish standard contracts that apply to many different situations. These contracts have been tried and tested, and you should become thoroughly familiar with them. You can also ask your current employer to show you the contracts he uses with independent contractors and consultants.

If and when you've satisfied yourself that these standard contracts do not adequately apply to your type of consulting, you should hire an attorney to assist you in putting together your own standard contracts.

Become familiar with enough of the contract basics so that you know when you're entering into a contractual agreement with another businessman. Remember, verbal contracts can be just as binding as a written contract. You must understand some of the fundamentals of contract law so that you don't wind up making legal agreements in situations where "it sounded just like a conversation."

Suggestion 4: Learn About Insurance and Agents

There are many types of insurance with which you should become familiar. Obtaining a good insurance agent is important; don't accept the first insurance agent who comes along. Like any other specialty, insurance agents have their competent and incompetent members. Ideally you should learn a bit about the area yourself. Insurance is not a difficult area to understand, and if you apply yourself and you use common sense, you won't have any trouble.

Proper insurance coverage should not be considered lightly. If you've been an employee for a number of years, you might have gotten to the point where you take a lot of your insurance coverage for granted. When you're going into business for yourself, you must take insurance as a new responsibility.

Group Coverage

You can obtain excellent group coverage through professional societies. The National Society for Professional Engineers, for example, has an excellent program. It covers major medical, life insurance, death and dismemberment insurance, disability income protection and many other forms of insurance.

Liability Insurance

The frequency with which consultants have been sued over the past twenty years has increased alarmingly. As a result, professional liability insurance has become mandatory in many circumstances. There are many types of technical consulting that cannot be entered into without a proper professional liability insurance program.

The type of liability insurance that you'll require will be strongly dependent on the type of consulting you perform. One of the quickest ways to find out what kind of professional liability insurance is required is to speak with professional societies and ask other consultants who are practicing in areas that are similar to yours. Be thorough and find out just what kind of coverage you'll need. Remember that even if you do excellent work and very rarely make mistakes, you could still be involved in a "shotgun" suit which includes everyone on the construction project.

Professional liability insurance can be expensive, but in certain cases it is necessary, and it's a cost that you will just have to include as an operating expense. There are some types of consulting firms that cannot operate without it, and you shouldn't take unnecessary chances. This is a situation where you will have to do a considerable amount of homework to find out what kind of coverage you need; it is not a situation where you should just take the advice of an insurance salesman.

The National Society for Professional Engineers and the American Consulting Council have taken considerable responsibility in professional liability insurance for their members. I suggest that you contact the professional societies in your field and obtain all their information concerning professional liability insurance.

Worker's Compensation

If you're planning on hiring employees, you'll be required to carry worker's compensation insurance. Worker's compensation insurance protects you as an employer from any claims that would be made against you by an employee during his employment. Your insurance agent will be able to explain the details. You should be careful to go over the state requirements and make sure that you are meeting them. Because the government is involved with this form of insurance, it tends to become cumbersome and involved. Note that there are different types of worker's compensation insurance. Some insurance is offered by private companies, and other types are obtained by funds that are backed by the state government.

Disability Insurance

If you have a sufficient bundle stashed in the bank, you'll be able to withstand any long periods of unemployment due to sickness or disability. When you're working for yourself, you must remember that if something happens to you, the money will stop rolling in. If you don't have sufficient funds in the bank, you should consider obtaining disability or salary continuation insurance. This insurance can be obtained either separately or in plans that are offered by professional societies. If obtained privately, it can be quite expensive, and you should compare what it has to offer to what could be arranged by using your own funds that you have saved specifically for this purpose.

Other forms of insurance which you should investigate are: general liability, automobile, travel, theft and damage.

Suggestion 5: Get All Those Licenses and Permits Handled

You're already familiar with the fact that Uncle Sam will try to get every penny out of you through taxes. Well, local and state government will also do the same with regard to licenses and permits. When you open a business, there are a few government agencies which will collect small sums of money from you. There are forms to fill out, fees to pay, licenses to obtain and regulations to meet.

The regulations and permits vary from city to city, and you'll have to take the responsibility of determining what is required in your locality. There are a few licenses and permits that are quite common in most areas. Your city clerk can give you the information.

Fictitious Name

You must obtain what is called a "Fictitious Name Statement." In some areas this is referred to as a "DBA" or a "Doing Business As" statement. You can obtain an application for a fictitious name statement through a town clerk or, since you are required by law to publicly advertise the fact that you're going into business, through your local newspaper. These fictitious names must be renewed periodically, usually every five years. If the name of your new company is unique, and you want to hold on to it, you should recognize that the renewal is important. Believe it or not—there are some people who make a practice of stealing fictitious names that have not been renewed and then selling the name back to the owner.

If you use your personal name as the official business name, the fictitious name requirement can probably be avoided. Check with your local town clerk to get the answer on this. If you're going to incorporate, you won't require a fictitious name statement; the name will be registered with the state through your incorporation procedures.

Business License

You'll also have to obtain a business license from the local authorities. These business licenses can vary in price from twenty to one hundred dollars. They have to be renewed every six or twelve months, and you can be sure that your local authorities will be happy to remind you that the bill is due. Your City Hall will be glad to give you the details.

Zoning Laws

If you are going to open an office, you should find out if your operation conforms with the local zoning laws. Usually, consulting operations don't have to worry about these factors, but if you'll be storing or using large pieces of equipment, you should inquire about the building codes and zoning requirements.

Seller's Permit

If, in addition to consulting, you plan on selling any items as products, you'll be subject to state sales tax laws and you might be required to obtain a seller's permit. You are required to collect sales tax from your customers and then transfer that tax to the state. This will be required unless you live in Alaska, Delaware, Montana, New Hampshire or Oregon. Whether you like it or not, you'll have to work for the state and play the role of tax collector.

Federal Identification

You have to register with the Federal Government as a business and obtain a Federal Identification Number. This identification number is used for identification purposes on licenses and tax forms. In the beginning you can use your social security number as a substitute. Your accountant can give you the appropriate IRS Form No. SS-5. If you're going to be hiring employees, you also will be required to obtain an employer identification number. This is obtained on IRS Form No. SS-4.

Suggestion 6: Improve Your Communication Skills

Improving your communication skills is a never-ending process. Communication is one of the keys to opening the door to a successful consulting firm. There are three different areas in which you can improve your communication skills considerably before you go into your own business.

Personal Communication

Excellence in inter-personal communication is probably the most important facet of becoming successful as a consultant. When you're dealing with clients, you're going to have to be able to pay complete attention to what they're telling you, and you must also be able to get your message across to them and make sure that they understand what you're saying. All the experience and expertise in the world will not help you succeed in consulting if your communication skills are weak.

When I first began consulting several years ago, I realized that my ability to communicate effectively was not nearly as strong as it should have been. I did some research into this area and found out that there were surprisingly few effective courses available that taught the fundamentals of communication. However, I did find one that merits attention. The Church of Scientology offers a course called The Success Through Communication Course, which has helped businessmen throughout the world dramatically improve their personal lives as well as their business potential. After taking this short, inexpensive course, my communication skills and confidence rose to a point where I noticed that my ability to run my consulting business and interact with clients had improved markedly. I've seen other people experience similar drastic improvements by doing this course.

Effective Presentations

If your type of consulting requires that you give presentations or seminars, you will want to improve your abilities along these lines. The most important point to remember with regard to presentations is that you need practice. In Chapter Seven we covered the ins and outs of presentations, and this should be very helpful to you. However, for the information to be of any value, you must practice with it and apply it when at the podium.

Take every opportunity you can find to give presentations on your present job. If your present position is such that you don't get an opportunity to give presentations, then get together with the marketeers in your company and ask them if you can assist them on a part-time basis.

If you've had little experience in speaking before a group, don't worry about all the fine details. Just get up in front of a group and start talking. Don't worry about how effective you are or about the improvement techniques that I mention in Chapter Seven. If you get practice in speaking in front of a group, you will realize that it's not much different than speaking to an individual. What really counts is getting a large number of hours of experience under your belt. If there are any original butterflies, they'll eventually die out.

If you don't have an opportunity to perform any public speaking within your company, then you might want to join a group which specializes in improving these important business skills. One such group is the "Toastmasters." There's probably a Toastmaster's group in your vicinity. You can find out by writing to them at Toastmasters International, Inc., 2200 N. Grand Avenue, P.O. Box 10400, Santa Ana, California 92711. The Toastmasters also has a magazine which features many excellent articles that will help anyone improve his public speaking skills.

Technical and Business Writing

Another area in which you can constantly improve your communication is your technical and business writing. Many technical professionals don't care to put a lot of their work in writing. However, as a consultant, you can improve your efficiency, effectiveness and communication by constantly communicating in writing. Again, if you want to improve your writing, just get into the habit of putting as many communications as possible into writing. Join in with your company when writing proposals and final reports and get as much experience as you can, keeping in mind that when you're on your own you might be responsible for 90% of everything that your company writes. Even if you go into a partnership or start working with a large number of individuals, you will still be responsible for evaluating the writing skills and abilities of others. Reviewing your basic grammar won't hurt one bit, but keep in mind that one of the biggest mistakes people make when it comes to writing is that they're so concerned about grammar that they often get little or nothing written. In the beginning, don't worry about your grammar or how it is written. Just write, and then, when that's over, write some more.

An excellent set of grammar books is entitled *Practical English*, and is published by the Career Institute, Inc., Sherman Turnpike, Danbury Conn. 06816.

Suggestion 7: Find the Right Secretary

In the early stages of setting up your business, you'll find a need for a very competent secretary. During the preparation period you'll be writing a large number of letters, and you will have to prepare yourself for being able to have technical reports and memos written exactly as you want them. The time to start looking for the proper secretary is right now.

Don't hesitate to start looking for a secretary because you don't have enough work for her/him. As you continue in setting up your business, you'll have more and more work to be done, and you must have somebody ready to do it. There are many excellent secretaries who are anxious to make a few extra dollars working after hours. Furthermore, there are many independent secretaries who run private secretarial services and will take on part-time work.

Locate a few independents and visit them in their homes or their offices and take a look at their abilities and their facilities. Your secretary is one of your most important employees. Choose her/him carefully. Some of the features to look for are as follows:

Typing and Dictation

Determine her/his capability and willingness to do technical typing and take technical dictation. I suggest hiring a secretary who has experience in this area; it's of great importance that she/he handle a large amount of your technical work with ease. There are a number of secretaries who have worked for technical organizations for many years, and they can reproduce much of your technical dictation without making a large number of spelling errors. Many secretaries who do not have experience with technical dictation listen to the technical words phonetically, and you'll wind up having to do draft after draft.

If you're going to be in business for yourself you're going to be doing a tremendous amount of correspondence, and your capability of producing a large number of letters in a short period of time will be a necessity. Dictating is not the whole answer. You must have a secretary who can understand what you're saying, reproduce it with correct spelling, and introduce grammatical corrections to your work.

Pickup and Delivery

Determine if she/he is willing to pick up and deliver the work that you will have for her/him. In the beginning, you will be working out of your home, and she/he will either work at an office or at her/his facilities which she/he has set up at home. You will not have the time to pick up or deliver materials which must be typed.

Compatible Hours

Determine what her/his hours are and if she/he is available in the evenings and on the weekends. While you're setting up your business, you are going to be busy during the day on your own job, and you are going to have to have somebody with whom you can work during your off-job hours. You might be able to find a secretary who works as an employee during the day and who is looking for some extra income.

Will She/He Take on Intermittent Jobs?

In the beginning you will find that there will be times when you have a lot of work for her/him, but you will also go through periods where you are working on other tasks and you won't be able to deliver any work to her/him. If this is incompatible with her/his objectives, you'll have to find someone who is more adaptable to your scheduling.

Typewriter

Determine what type of typewriter she/he has available. At present, for a combination of technical and regular typing, I can only suggest an IBM Selectric—preferably the kind which has the self-correcting unit. If you have found a secretary who seems quite compatible but doesn't have access to an IBM Selectric, consider renting or buying one.

Transcriber

Determine what type of dictation transcriber she/he has. Make sure that it's compatible with your own dictation equipment and with the tapes you'll be giving her/him.

Photocopier

Determine if she/he has, or has access to a photocopy machine. It seems today's office activity can't avoid a tremendous amount of duplicating.

Technical Reproduction

Determine if she/he has had any experience producing technical graphs and charts. If she's/he's had experience doing technical work, the chances are she/he might be experienced in reproducing technical illustrations. You're going to need some assistance in reproducing technical information, and if you can find a secretary who handles this job, you'll save yourself a considerable amount of time. Ask her/him if she/he knows how to make view-graph transparencies, and if the photocopy machine which she/he has access to will accommodate them.

Pay Scale

Determine how she/he wants to get paid. In the early stages of your work, you certainly don't want to hire an employee. Ideally, she/he should invoice you for the hours that she/he works, any mileage that she/he uses in delivering your work, travel time, and other expenses. Agree on a fair pay scale, and don't forget to reimburse her/him for all her expenses.

Library Research

Determine if she/he is capable of and willing to do some library research for you. Library research can be a lot of fun and many secretaries look forward to it. It will be of great assistance to you if you can have someone who can find and deliver library information. You want to do everything you can to optimize your time. Having somebody who is capable in using a library can be very beneficial. Determine if she's/he's had experience with this. There are many people who have been inside a library, but who don't really know how to use the facilities. If she/he does not know how to use them, consider bringing her/him to the library and introducing her/him to the facilities that you use most often.

Bookkeeping

Determine if she/he has any experience in bookkeeping. She/he doesn't have to be an accountant, but if she/he knows the basic elements of keeping books, you might want her/him to keep the business records for you. Even if she/he has little experience in this area, you can train her/him to perform the necessary tasks. Many secretaries have experience in paying bills and keeping the checkbook balanced.

When you find somebody who meets a large number of the above requirements, give her/him some technical work to reproduce and judge its quality. Try this out with three or four different secretaries and pick the one you find who will be the most compatible with your work and schedule. You must remember that when you get rolling, your secretary is going to be an extremely important part of your business operation. She/He should be treated with the respect of a professional who takes pride in her/his work.

Suggestion 8: There's Fun and Profit in Being a Dictator

The first draft for this book was dictated in a few weeks. A few years ago it would have taken me an entire evening to draft and get a business letter typed. Now, with the same skills which you will have soon, I can produce 100 typewritten pages in three or four hours of dictation.

You want to be in business for yourself and, no doubt, you want time for yourself and your family. Extra time is dependent on your ability to communicate. You will have all kinds of people to whom you must communicate: lawyers, accountants, potential customers, clients. How are you going to do it?

If you can dictate, you can communicate rapidly and effectively. I'm not only talking about letters to people. I'm talking about all forms of communication, even reminders to yourself.

Here's a list of a few places where the use of a dictating machine comes in handy:

1. Taking notes in the field and "on the site."
2. Noting important points after a business meeting.
3. Initiating quick memos to your associates as soon as they come to mind.
4. Initiating letters to prospects.
5. Answering large quantities of mail in just a few minutes.
6. Writing first drafts of entire reports. These first drafts can then be given to your working associates who will finish the work that you have started.
7. Follow-up letters to telephone calls.
8. Personal letters to friends and associates. Some people say they would rather write a personal letter longhand, but if I have a choice between dictating a letter and no letter at all, I'll go with the typed letter.
9. Making long lists of items that have to be acomplished by others. An example would be a list of things that an auto mechanic should look into to have your car brought into its proper condition. How many times have you driven from a garage and realized that you

forgot to tell the mechanic to do an important task? Sometimes writing a list at the last minute is not the most efficient.

Increasing Production and Gaining More Time

Each one of us communicates every day of our lives. But for most of us, the quality and volume of our communication doesn't increase very much. One of the ways that quality and volume can increase is through the use of a dictating machine. Just think of all the time that can be wasted if you have 15 or 20 phone calls to make in a day. Often the people you call aren't there, and you've wasted a tremendous amount of time trying to get through to them. When they return your call, you might not be available. The habit of writing short letters to get your brief message across not only saves you time, but it also allows you to say what's important and then get onto more important tasks. Additionally, it gets your orders into a written form so there will be no misinterpretation as to what you said.

The efficient use of a dictating machine can add an extra hour or two to your business day. That might sound incredible to you, but it's true. Your efficiency increases to the point where your output is considerably more than it used to be.

A very interesting thing also happens when you become proficient in your dictation. The quality of your work changes somewhat. Part of the reason for this is that you don't spend a lot of your time and energy thinking about whether something is said correctly. Especially in the beginning, don't worry about how you say things; just say it. After a while your conversational accuracy will improve and you will get your points across effectively.

Letter Writing

In the beginning, don't tackle formal letter writing with your dictating machine. Just pick a friend or two with whom you'd like to get into closer communication and write them, explaining that you're trying to learn how to use your dictating machine and that you've chosen them to suffer through with your dictation. They'll enjoy hearing from you, and might even appreciate not having to unscramble your handwriting.

Continue dictating personal letters until you get to a point that is a very important transition stage in dictating. You will come to a point where you no longer think about how you're saying things; instead, you'll find yourself just speaking. It will be very similar to what happens when you're in a conversation. In a conversation you don't find yourself straining about how to say something; you just say whatever you have to say without thinking about it. Some friends enjoy receiving the actual tape from you. It's a very personal way to communicate. Try surprising a friend by sending him a taped message. He'll enjoy it, and it will save you time.

Suggestion 9: Setting Up the Office

While you are working on your present job, you should set up an office at home. In the beginning this home office can be used as your regular business office. Until you get a substantial income coming in, you should not rent a larger office space and spend a lot of money on fancy furnishings. Just concentrate on the fundamentals.

In setting up a home office, you should mark off a portion of your home exclusively for your business purposes. Get a very large desk with a comfortable chair. You're going to be spending a lot of time at this desk, and you'll want it to be as effective and efficient a tool as possible.

Filing

Invest in a good four-drawer filing cabinet. Spend some time setting up an effective filing system which can be used to retrieve information *rapidly*. If your filing system is set up correctly, you should not have to do any of the filing yourself. You can have your secretary or one of the members of your family do it for you. The filing system should be cross-referenced, and whenever something has to be filed, you can jot a shorthand notation on the document stating where it belongs. Then, throw it in your "out" basket and it will be filed for you. This might sound rather formal, but it is important, because it will save you a lot of time.

Invoicing

The type of invoicing that you'll be doing will depend on the type of consulting in which you're engaged. Writing a business invoice is important because it's your direct communication line with the necessary individuals who are responsible for paying you.

Often contractors omit certain important details when they invoice their clients, causing unnecessary problems. Be sure all the necessary data is on the invoices you deliver. If you will be doing any government contracting or subcontracting work, you'll have to obtain the particular information that is related to invoicing government-related work.

Typewriter

If your secretary does not have access to an IBM Selectric typewriter, you might consider buying a used one. They can be purchased in a refurbished condition for $700 or $800. If you're going to be performing professional consulting, you'll be in need of this kind of typewriter. Unless you can find a brand that delivers similar quality with a high versatility, you shouldn't consider any other type. This especially holds for consultants who will be writing technical reports. This type of typewriter is also excellent for including mathematical symbols and Greek letters. There are some types of engineering consulting work in which this type of typewriter is indispensable.

When you're purchasing your machine, make sure you get a service warranty with it. These service warranties will guarantee that your machine is in working order at all times. Don't allow yourself to be caught in the situation where some of your important equipment breaks down on the very day that you're supposed to get an important proposal or report out the door. There is a charge for this service warranty, but this is one area where you shouldn't try to save a few dollars. These are sophisticated pieces of equipment, and they often have minor breakdowns.

Dictation Equipment

You should have a reliable cassette dictaphone and transcriber. These should be purchased after you have found a secretary. In some cases she'll/he'll have a transcriber, and you might want to buy a dictaphone that uses cassettes which match her/his transcriber. Have a full supply of 15-, 30- and 60-minute tapes available, and get used to performing a lot of your work through dictation.

Computer

You might be in the need of a home or office computer. If this is the case, you should not wait until you start your business to buy one. Sometimes buying the correct computer requires a bit of shopping; there are a number of excellent brands on the market. If you do get a computer, make sure you get one which has some form of paper printout mechanism. The investment for a printout is not large when compared with the cost of the computer, and it will save you considerable time in your computations.

When you're in business for yourself, you should be spending as much time as possible with your clients or generating new business. If you spend too much of your time trying to debug computer programs instead of debugging a faulty sales program, you will find yourself in the soup. Therefore, get a machine that is easy to use and is reliable.

Photocopies

If you anticipate a lot of photocopy work, you should check into picking up a used photocopy machine. There are some small machines on the market today for under a thousand dollars. If you don't get a photocopy machine, locate one that's in your vicinity to which you will have easy access. It's pretty difficult to operate an office without access to a photocopy machine, so you should make sure that you have a high-quality machine available to you at all times.

Phone

Don't use your home phone for your business purposes. For tax purposes, you have to keep a separate record of your business calls.

Another important but often overlooked point about the telephone is that it is beneficial to have two business numbers—one number is the one you advertise and receive incoming calls on. This is the one to which you connect your answering service or machine. The other number can be used for outgoing calls. If you are in the type of business where you are on the phone a large portion of the day, having two numbers will enable you to maximize your communication with your clients and business associates.

Answering Service

Make sure you get an answering service. The service usually costs between $25 and $40 a month, and it's worth every penny of it. Under no circumstances should you have a phone system that's not being answered 24 hours a day. Sometimes people use tape recorders and answering machines to handle incoming calls when they're not in the office. This system is less expensive, but it is certainly not as efficient. There are some people who just don't like talking to a tape recorder and, no matter how sophisticated you make your system, it won't be the same as an individual answering the phone for you. The major advantage of the answering machines is that they record the entire message accurately. Answering services can be rather inaccurate in recording the details of messages.

Technical Equipment

Finally, you should do a complete survey and determine all the technical equipment you'll be needing for your operation. Make a full list and then determine what equipment you can buy used, and what must be purchased new. Don't go overboard on equipment that you might not need. Just get what is necessary to get you started. You can always pick up fancier equipment after you've won a few contracts.

Suggestion 10: Promote and Advertise

If you want to get a solid start as a consultant, you're going to have to promote heavily and advertise your capabilities. There are many ways you can do this before you even start your own company.

Write Technical Articles

One of the most effective techniques is to write a large number of technical magazine articles. These articles will get you into communication with potential clients throughout the country. This can be done to your advantage while you are still employed.

Publishing

Another technique is to promote your technical expertise by writing papers for technical journals. No doubt you've probably already written some papers and have found that individuals you didn't know were very interested in what you had to say.

Once you have written a few magazine and technical journal articles, you then have a basis on which to promote. When you finally leave your present position and go on your own, you should have a long list of people to whom you can send your reprints. Don't ever assume that your public has read your articles. If they have and you send them a copy, they can always throw it away. But, the chances are that they are not familiar with your technical expertise, and you should have a mail campaign that is constantly mailing your articles. If you have written a total of six such articles, you might find that sending one reprint every two months to prospective clients is quite effective. Include a short note with each article, and mention that you thought the article might be beneficial to their work. They'll appreciate the fact that you have thought of them, and they might be able to use the information that you send them.

Brochures, Resumés, Stationery

While you're working, you can also prepare a consulting brochure or resumé which summarizes your abilities.

Don't forget to get your stationery, business cards and other office papers printed. In a sense, these items are also promotion materials, and getting them prepared before you go on your own will save you a considerable amount of time that can be spent working on contracts.

Suggestion 11: Line Up Some Clients

You don't go into business by yourself by hanging your shingle on your door and waiting for somebody to come along and tell you that they need your services. You've got to go out and make your services known.

If you plan to stay on your present job for at least six months or a year, you'll have an opportunity to get into communication with a large number of *individuals* who are potential clients. Note that I'm saying that you are to get in touch with individuals and not companies. It's true that you're paid by a client company, but remember that it will be an individual within the client company who is your point of contact.

Generate a List

One important task is to generate a list of every possible client and every individual who could put you in contact with a potential client. The formation of this list should be an on-going task. Write down as many names, addresses and telephone numbers as you can. Then, over the next six months to a year, add two or three names to the list daily. You should set a quota every day of how many names you could add to the list. This list will be one of the most important documents in your entire business. After you've generated a list of one or two hundred names, you should review the list and identify those prospects whom you believe have the most potential for becoming clients. Get in contact with each one and arrange a luncheon engagement to discuss the possibility of your working as a consultant and getting contracts through them.

If it appears that you will not be able to get a consulting contract with them, ask them if they know of anyone else who could use your skills. Usually the individuals who can't hire you immediately will know someone either in their company or elsewhere who might be able to use you.

Don't Jump the Gun

If you have done your homework correctly and you've contacted enough individuals, you might be surprised to get one or two immediate offers. Don't jump the gun, however, and leave your job prematurely. Just continue to establish as many potential clients as possible. Plan on speaking to at least one potential client a day and get his opinion about hiring you as a consultant. When you are speaking to these prospective clients, determine what their needs are and what type of specialist they are looking for. You might find that there are some needs that are closely related to your abilities, and that by taking a few brushup courses or reviewing in a new area, you might be able to obtain consulting jobs in subject areas where you haven't considered yourself an expert. Remember, if a demand is high enough, many companies will be willing to give you a try. If you go in and find you can't do a job, nothing will be lost. You can inform your client that you're not the proper person to do the work. You should not fear to give many jobs a try, and you will probably find that you can handle many that initially appeared as if they were not within your abilities.

Suggestion 12: Put Your Finances in Order

Make a listing of all your personal debts and outstanding loans and set up a program to pay them off. Not owing anybody a dime and having some extra money in the bank is the ideal way to go into business. Since you'll be generating enough debts by starting up your own business, you should set up a program to take some of the money you're earning on your present job and pay your bills.

Savings

After you have your bills paid, set up a savings plan in your company name and put in regular contributions which will be later applied to setting up your business. More important than the amount of money you put into this plan is the simple fact that you start a plan and contribute to it on a weekly or bi-weekly basis. No matter how small, get something started. Even if you're only putting in $10 or $20 a week, you should make sure that something goes in on a regular basis. This will be one of your first business decisions, and if you don't save money on a regular basis, nothing's going to grow.

Checking Account

Start a business checking account. There will be many items that you'll be buying that are directly related to the business. You will want to keep these items on record as being bought for your business. They're part of your initial business investment and can be written off on taxes at the end of the year. Get the checking account put in your name. If your secretary will be handling your banking, then make sure that you set up your account so that she can deposit and withdraw funds.

Raising Capital

It has been said that it's more difficult to raise capital as a sole proprietor than to raise capital as a partnership. This is a generality that doesn't really mean much when it comes down to your personal situation. The amount of capital that you can raise is determined solely by how well you can do it!

If you find a partner who has the ability to raise capital, you can raise more through his abilities, but don't kid yourself. Prior to forming a partnership, you should know some of the ins and outs of raising capital and financing. This does not mean that you should refrain from forming a partnership. However, if you are forming a partnership because you really don't know how to raise a plug nickel, you might be headed for trouble. The answer lies in learning how to raise capital and then raising it.

Make a careful review of your financial situation and determine how much money you can raise for starting your new venture. Go over your financial scene thoroughly with your family and make sure there are no financial conflicts. Make a list of all the equipment that is absolutely necessary and other initial operating costs. This should not run more than a few thousand dollars, and in some cases it will be under a thousand. If at all possible, do not borrow the money to start your consulting venture.

Accountant

Visit a number of accountants after you have reviewed some finance basics. Determine the best method for keeping your books and inquire about ways that you can improve your credit. Decide on working with one of these accountants and stay in close communication with him. Don't hesitate to pay him for his time; he'll be well worth the money. He will also be able to help you with the pros and cons related to operating as a sole proprietorship, partnership or corporation.

Suggestion 13: Establish Credit

Visit a few bankers and tell them that you're planning on going into business for yourself in six months or a year, and talk to them about establishing a business credit rating. It is very important to establish a personal relationship with one or two bankers who will have faith in your operation, because it is your personal credit rating that is going to form the foundation for any business credit that you'll obtain in the future. You won't be able to obtain a business credit rating until you've been in business for a year or more. Therefore, your personal credit rating is very important.

Also, find out what will happen to your personal credit rating if you leave your job and go to work on your own. There are some banks that will change your credit rating due to the fact that you no longer have a regular established income.

The important point with regard to the banks is not only a numerical credit rating, but also the establishment of a solid relationship with the bankers. Visit a few bankers and then determine which bank officers you get along with the best. Get to know them well and visit them often. Show them that you're a productive businessman and that you won't be a high risk in case you have to obtain a loan.

Discuss with the bankers what kind of plans you can initiate to increase your credit rating. There's no reason why you couldn't bring your credit rating into the $10,000, $20,000 or $30,000 range rapidly. If you have to, borrow money from the bank and pay the loan off with the money that they've lent you. This will cost you some money, because you'll be borrowing a few thousand dollars, and there will, of course, be interest, but it might be the only means you have to increase your rating.

If you've had a bad credit history, you'll have to prove to the banks that you're a good risk. The most efficient way of doing this is by paying your bills on time. Get involved with all the creditors with whom you've had trouble and ask them what is necessary to improve your credit rating. In most cases, it'll just be a matter of getting your bills paid regularly for a year or two.

Suggestion 14: Investigate Buy-Sell Agreements

Life Insurance for the Sole Proprietor

If, after looking into a partnership and a corporation, you've decided to run your consulting practice as a sole proprietorship, then you should consider looking into your present life insurance to make sure that it is sufficient to handle your family in the case of your demise.

Don't make the mistake of thinking that as a sole proprietor your business debts will be separate from your personal debts. The law states that no distinction is made between your sole proprietorship and your personal estate. Therefore, if you die, your family may find itself in a rather severe situation. If your business is in debt, your family will be responsible to pay those debts if you die. To avoid this, you should have sufficient life insurance.

Also, you should recognize that there aren't that many options for a sole proprietorship to continue if you pass on. One of the options is for the entire business to be liquidated. In this case, any debts would have to be paid by your family. Another option is that your family or heirs could step in and continue running the business. Now in many businesses this might be fairly straightforward, especially if you are running a family grocery store or something of this sort. However, in the case of technical consulting, it might not be too simple for your wife and kids to instantly become consultants. Therefore, the option of their coming in to run the business might be a very impractical route to go.

In a case such as this, a sole proprietorship could be sold to one of your employees or an outsider. If this possibility exists in your situation, a Buy-Sell Agreement can be set up right now with your attorney, and it can stipulate that the executor to your estate will sell the firm to the appointed person or persons at a fair price, which is either set down by formula or is periodically changed in the Buy-Sell agreement.

If you choose to go the route of the Buy-Sell Agreement, you should make sure that the person who buys your firm has enough money to do so. If he doesn't have money to buy you out, then you should definitely purchase a business life insurance policy that will guarantee the financing that will pay for your firm.

With this type of Buy-Sell Agreement, the executor to your estate is legally bound to sell your business to the person who has made the Buy-Sell Agreement with you. In this way, your family is guaranteed that they will have no debts to pay, and instead, they will receive a fair price for the business. Business life insurance for a sole proprietor is a fairly common type of insurance, and you can get more details about this kind of insurance from your insurance agent or your attorney.

Buy-Sell Agreements for Partnerships

If you are planning on starting your consulting practice with the formation of a partnership, then there is a simple agreement that you could make that could save a lot of grief in case one of the partners dies. Consider the following situation:

Bill and Frank were two good friends, and they have worked as mechanical engineers for about twelve years. They decide to start their own consulting practice, and after a couple of years the practice is really booming, and they have enough income to easily support their two families and expand into other areas. It's one of these ideal partnerships where one of the engineers, Bill, takes on most of the technical aspects of the work, and his partner takes on all of the administrative and sales work. Suddenly, in a plane crash, Bill dies and Frank is forced to liquidate the entire business, even though the business is doing quite well and is providing his family and Bill's family with a substantial income.

Frank really has no choice in the matter. If you remember from Chapter 14, the law states that a business partnership ceases to exist upon the death of one of the partners, unless a specific agreement was made such that the business would continue.

Now all Frank had to do to continue the business was to get Bill's wife and family to agree to create a new business relationship and carry on the old business as usual. However, Bill's wife and his children had no interest in keeping the old business going. Therefore, Frank had no choice but to sell all the assets of the firm, pay Bill's family their share in cash, and then try to start all over by himself.

There is a way to avoid a situation like this. If proper arrangements had been made in the beginning, Frank could have bought out Bill's family's interest, in cash, and could have become the sole owner of the company. This certainly would have been a more pleasant ending. In fact, with all due respect to Bill, it would have been just the start of a new phase for the company. Bill and Frank could have entered into a Buy-Sell Agreement, stipulating that the survivor of the partnership will buy out the deceased partner's interest. Also, the heirs are obliged to sell their share of the company to the partner. The Buy-Sell Agreement, which is drawn up by an attorney, has the agreed upon price, or a price formula stipulated so there will be no problems concerning the value of the business.

Suggestion 15: Get to Know Other Consultants and Placement Firms

Don't start by operating in the dark. Find out who some consultants are in your specialty and get to know them. When they're working late, go to their office and chat with them for a few minutes. If you can, help them out a little bit. Get to know what their operations are like, and you'll find that there's no mystery to running a consulting operation. Getting some "hands on" experience and witnessing a few consulting practices in operation will show you that consulting is a possible reality. It might be a lot of hard work, but being on your own is a very real thing, and it will only become real to you if you start to witness it. Find some consultants to work with on a part-time basis. It can be an excellent means for making a few extra dollars and gaining the valuable experience of interacting with technical professionals who have already gained some business know-how and independence.

When you're taking a close look at these consultants, study them carefully and observe their strong and weak points. Take notes and try to decide what they're doing right as well as wrong. Observe their operation closely and recognize that many of the tasks that they have to perform are very similar to what you're already doing as an employee. Once you see that there really are many similarities, you'll begin to recognize that working as a consultant is rather straightforward. One of the major differences, however, is the fact that when you're on your own, you'll be taking on many new responsibilities. When you close up at night, it won't be the same as walking out the door of your plant and going home to mow your lawn. You have more responsibility, and you might find yourself working late evenings and having a lot more to handle. With that extra responsibility, though, comes more control and independence, and that's where a lot of the fun really lies.

Suggestion 16: Consider Becoming a Registered Professional Engineer

Besides those business forms and licenses already mentioned, there is no license required for most consulting engineers. However, in many states, before you may work on projects that affect public welfare, you must become a registered professional engineer. You should inquire about the requirements of becoming a registered professional engineer.

There are many types of consulting practices that don't require that you be a registered engineer. For example, you might work on electronic engineering projects that were not directly related to the public. In any case, you should inquire with your state officials to determine if you must be a registered engineer.

You'll be required to pass a state exam to obtain your registered engineer's license. It's helpful to study for these exams because many of the questions might not relate directly to your specialty. More information can be obtained in the excellent book, *Professional Engineer's License Guide*, by J. D. Eckard, P.E., (Herman Publishing Inc., 45 Newbury St., Boston, Massachusetts 02116).

When you go to register as a professional engineer, you are not to register as a partnership or a corporation. You are to register as an individual, no matter what form your organization takes. The registration laws of most states require that if you're going to be dealing on projects that are related to the public welfare, then you must notify the state in which you are acting in an engineering capacity. If you are in a type of engineering where the states require that you obtain a professional engineering license, you should make sure that your company is controlled by someone with a proper license. If you have a partnership, at least one of the partners should be licensed. In a sole proprietorship the owner must take this responsibility.

Suggestion 17: Get Your Business Operation Forms

Depending on the type of consulting operation you'll run, you'll need some standard business operation forms. If you plan on consulting in areas related to the construction industry, these forms can be many. You can design many of your own forms fitted to your own needs. If you're an engineer, I suggest you get in contact with the National Society for Professional Engineers or the American Consulting Engineering Council and get some of their standard forms and contracts. Examples of forms which you might need are:

Change Orders
Daily Progress Reports
Fee Schedules
Agenda Records
Time Cards
Construction Progress Reports
Expense Records
Internal Memo Forms
Statements for Payment and Invoices

You should make a list of the forms that will be required for your specific applications and get copies of them so you will have them when you get ready to go to work.

Suggestion 18: Set Up Your Consulting Library

Make a collection of texts for your reference in finance, accounting, law, insurance, and any texts that will help you out in your specialty. Once you're on your own, you're not going to have access to a company library. If you don't have a sufficient library of your own, you should make the investment to make sure that you have all the research and reference material so you can deliver an excellent service.

Go to your local libraries and study what's available with regard to books that can help you out in your specialty or any specialty business matters. Make a full listing of all the books that are available which could be of use to you and keep this record with you. Some libraries will let you take the identification cards out of the card catalog, and you can then photocopy them and use the photocopies for quick reference in your office. Buy any necessary books or reports which will assist you in forming a useful library.

When it comes to deciding on information for your business, you should recognize that the value of information is dependent on how you use the information. Don't judge the cost of a book or a report by comparing it to other materials; judge it by its value to you. An example would be the value of this book which you're presently reading. For most people, the material in this book will save them considerable time and money. If you can find information which will save you time (time that would be required in trying to find the information on your own), then this information has considerable value. Be willing to pay for valuable information. Sometimes I'll pay $20 or $30 for a book or report and only use it two or three times. Often just one fact pays for the entire publication.

As a consultant, you're being paid not only for answers, but to get answers for your clients rapidly. It's your effectiveness and efficiency that is in demand. If there exists a good research library in your vicinity, you should get to know it thoroughly. Then, get someone who can do library research for you. Often you can find students at universities who will be happy to obtain information for you. All that's necessary is a telephone call to them and they'll go the library and pick up the necessary material. Your time is too valuable to be spent looking for research papers. Having someone who can get the information for you is more efficient and less expensive.

16
Further Help

In writing this book I have tried to include only that information which one could not readily find elsewhere. However, there exists a great deal of information written by others that can be of great assistance to you. The purpose of this chapter is to provide you with extra reference material that I have found helpful in running my own practice.

More On Taxes

If you have some fears about the IRS, I suggest you read *All You Need to Know About the IRS* by Paul N. Strassels (Random House, New York, 1979). Paul Strassels used to be employed as a tax law specialist in the IRS. His unique book shows how the officials in the IRS think, and how they create a strategy of fear to make people pay undue taxes. He also shows how the IRS uses their major weapon—the tax audit. The basic purpose of the tax audit is not necessarily to take more money from you, but to scare you and your friends into being honest about your taxes. I recommend this book to anybody who is planning to go into business for himself.

B. Ray Anderson's *How You Can Use Inflation to Beat the IRS* is a book that tells you how you can legally avoid paying taxes and protect your assets from the boys in the IRS. Once you're making money as a consultant you're going to want to keep it, and this book shows you how to use tax shelters and deferral techniques to hold on to your money. Extra ideas are also given for modifying your pension plan and using trusts. The book is published by Harper and Row, New York.

An excellent tax service that gives you monthly information on small business tax techniques is the *Tax Prompter*. Information on annual fees and other tax related services can be obtained from Dara Business Publications Company, Incorporated, P.O. Box 3158, Richmond, Virginia 23235.

An excellent text on tax strategy and tactics is *Taxation for Engineers and Technical Consultants*, by Mark J. Lane (John Wiley, New York, N.Y.,

1980). This book helps the technical professional with financial decision-making on day to day activities as well as strategies associated with mergers, acquisitions, and research and development issues.

Accounting for Technical Consultants

Although the area of accounting and bookkeeping is a very important one for a consultant, I have deliberately not put much emphasis on it in this book. This is because I recommend that every consultant get more detailed advice in this area than space would permit here. *Simplified Accounting for Engineering and Technical Consultants*, by Rick Stephan Hayes and C. Richard Baker (John Wiley, New York, N.Y., 1980), is a financial survival kit for the technical professional in private practice. It deals with accounting techniques that are specifically related to a consulting practice. If you have decided that consulting is for you, then one of your next steps should be the acquisition of this book.

The book consists of the following material as it relates to a technical type of practice: recording expenditures, billing clients, tax credits, recordkeeping, income statements, balance sheets, tax deductions, tax treatment of corporations, loans, depreciation, Subchapter S corporations, partnerships, wages, etc.

Publishing

There is no quicker way to become known in your field and get consulting contracts than to publish a book in your specialty. *A Writer's Guide to Book Publishing* by Richard Balkin (Hawthorn Books, Incorporated, 260 Madison Avenue, New York, New York 10016) is a book that will give you insight into how to work with publishers. This book is filled with advice aimed at helping would-be authors to better understand the game of publishing. Once your book is published you might find that you only make a few thousand dollars a year on royalties, but you could easily make twenty or thirty thousand dollars a year through the new business that is generated once your name is known to so many readers. This book will not guarantee every writer a publishing contract, but it will definitely give you the proper perspective that is needed prior to writing a manuscript.

If you're interested in copyrighting any technical reports, you should be familiar with *The Writer's Legal Guide* by Tad Crawford (Hawthorn Books, Incorporated, 260 Madison Avenue, New York, New York 10016). The topics covered in this book are the general rights of writers, the new copyright law, contracts, etc.

Raising Capital

If you need assistance in seeking venture capital for your consulting practice, I suggest you obtain a copy of Stanley Rubel's *Guide To Venture Capital Sources* (Capitol Publishing Corporation, 2 Laurel Avenue, P.O. Box 348, Wellesley Hills, Massachusetts 02181). Even if you're not interested in raising venture capital at this time, this book is still excellent for anyone who is starting a business and wants to know how the venture capital process works. This book is especially valuable to a consultant who works with companies who have to raise money. It's also of assistance to investment bankers, commercial bankers, lawyers, and accountants.

A complete summary of all the different institutions which can assist you and your financial needs is given in Dileep Rao's *Handbook Of Business Finance And Capital Sources* (InterFinance Corporation, 305 Foshay Tower, Minneapolis, Minnesota 55402). This book gives a very good introduction to the fundamentals of business finance, and then goes on to describe all the different ways of raising capital. It also gives a good description of all the different types of sources of capital such as commercial banks, commercial paper houses, commercial finance companies, factoring companies, investment bankers, pension funds, etc. Information on private financial institutions and different federal programs is also summarized.

If you're interested in borrowing money, I would suggest Rick Stephan Hayes' *Business Loans* (CBI Publishing Company Incorporated, 51 Sleeper Street, Boston, Massachusetts 02210). *Business Loans* reviews all the different sources of loans, both private and public, and then goes into detail showing a person how to prepare a loan proposal, do basic financial calculations, and present the final loan proposal.

Direct Marketing and Mail Order Techniques

If you're interested in promoting your practice using direct mail techniques, I would like to suggest Bob Stone's *Successful Direct Marketing Methods* (Crain Books, 740 Rush Street, Chicago, Illinois 60611). This book gives an excellent review of all the direct marketing techniques. It has a very good introduction to using mailing lists. I've used this book quite a bit myself and have found it very helpful. It also has chapters on magazine and newspaper advertising, as well as a chapter on telephone marketing.

Patents

While consulting you may come across many ideas that are worth money. If you're not sure what to do with these ideas, you might consider taking out a patent. David Pressman's *Patent It Yourself* (McGraw-Hill Book Company, New York, New York) is a book that shows you the ins and outs of patenting your invention yourself. He brings you through every necessary step in the patent procedure. This includes record-keeping, testing, evaluating, searching, patenting, and selling. This book is written by a knowledgeable practicing patent attorney, and it is packed with useful information that can be of great assistance to you.

Engineering Societies

The Engineers Joint Council (345 East 47th Street, New York, New York 10017) publishes the *Directory of Engineering Societies And Related Or-

ganizations. The Engineers Joint Council of the Federation of Engineering Societies has the objective of advancing the art and science of engineering in the public interest. This directory gives detailed information on over 460 different engineering and technical societies. Some of the societies that relate to consulting are:

ASSOCIATION OF CONSULTING CHEMISTS AND CHEMICAL ENGINEERS (ACCCE)
50 East 41st Street
New York, New York 10017
(212) 684-6255

ASSOCIATION OF CONSULTING MANAGEMENT ENGINEERS, INCORPORATED (ACME)
230 Park Avenue
New York, New York 10017
(212) 697-9693

AMERICAN CONSULTING ENGINEERS COUNCIL (ACEC)
1155 15th Street, N.W.,
Suite 713
Washington, D.C. 20005
(202) 296-5390

CONSULTING ENGINEERS ASSOCIATION OF CALIFORNIA (CEAC)
1308 Bayshore Highway,
Suite 101
Burlingame, California 94010
(415) 344-5782

CONSULTING ENGINEERS COUNCIL OF COLORADO (CECC)
111 South Colorado Boulevard,
Suite 305
Denver, Colorado 80222
(303) 757-3379

CONSULTING ENGINEERS COUNCIL OF MINNESOTA (CEC/MINN)
5009 Excelsior Boulevard
Suite 126
Minneapolis, Minnesota 55416
(612) 922-9696

CONSULTING ENGINEERS COUNCIL OF MINNESOTA (CEC/MINN)
5009 Excelsior Boulevard
Suite 126
Minneapolis, Minnesota 55416
(612) 922-9696.

CONSULTING ENGINEERS COUNCIL OF NEW YORK STATE, INCORPORATED (CEC/NYS)
3010 Troy Road
Schenectady, New York 12309
(518) 785-6833

CONSULTING ENGINEERS COUNCIL OF OREGON (CECO)
5430 S.W. 90th Court,
P.O. Box 25082
Portland, Oregon 97225
(503) 292-6808

CONSULTING ENGINEERS COUNCIL OF UTAH (CECU)
P.O. Box 11722
Salt Lake City, Utah 84147
(801) 531-8793

NATIONAL ASSOCIATION OF BLACK CONSULTING ENGINEERS (NABCE)
667 South Jackson Street
Seattle, Washington 98104
(206) 682-2641

NATIONAL SOCIETY OF PROFESSIONAL ENGINEERS (NSPE)
2029 K Street, N.W.
Washington, D.C. 20006
(202) 331-7020

INDEPENDENT CONSUL-
TANTS OF AMERICA (ICA)
567 W. Westfield
Indianapolis, Indiana 46308
(317) 251-8411

Professional Liability

In many areas of engineering, a consultant may be vulnerable to lawsuits. If you're not sufficiently covered by insurance you might have your practice interrupted or shot down. To minimize the possibility of a lawsuit against your practice, I suggest you invest in a course offered by the Management Institute, 2 East Avenue, Larchmont, New York 10538. The title of this course is *Reducing Professional Liability in Your Engineering Practice*. This course covers the following subject areas: liability related to design, liability owing to client requirements, shop-drawn approvals, construction phase liabilities, liabilities relating to contractors, liability for health and safety, and third-party liability. This short course is filled with many case studies taken from practical engineering situations. If you have questions about liability, this course is a good first step.

Consulting Publications

The *Consulting Opportunity Journal* is an excellent consulting newspaper that is published by Consultants Media, 1629 K Street, N.W., Department MKT, Washington, D.C. 20006. This is a quarterly newspaper that covers all aspects of consulting. It is written for technical people as well as consultants in other fields. In 1981 the subscription price was only $15 a year, which is a fantastic bargain. One of their issues covers such subjects as: conducting seminars, direct marketing, how much should you charge, how to get free advertising, negotiating your fee, selling your services, etc. I recommend this excellent newspaper to anybody who is either a consultant or who is just considering consulting as a new venture.

Handling Your Personal Fears

For most individuals, it is quite a big step to break off on their own. In counselling engineers and data processing professionals in starting their own practice, I have learned what it takes to succeed in starting one's own practice. Everyone I have ever counselled has had his own unique set of personal fears and "hang-ups" that have prevented him from expanding into new horizons. The difference between the winners and the losers is not whether or not they have personal fears and hang ups. We all have them. *What distinguishes the successful consultant from the unsuccessful one is his recognition of and attitude toward his fears, inabilities, and hang-ups.*

My counselling experience has proven to me that the successful person is highly aware of his personal hang-ups that could prevent him from achieving his goals. He is most often very willing to communicate the most intimate particulars of his idiosyncracies, and he looks upon his hang-ups as barriers to be overcome in the attainment of his goals. No matter how strange he feels his fears and hang-ups are, he is eager to do something about them rather than pretend they are not there. He sees the strong and weak sides of his character, and is anxious to correct the sides that are holding him back.

Consider, for example, the person who becomes petrified when he has to speak in front of a small group. He's very able technically and knows that he has the potential to double or triple his income if he becomes a consultant. He openly admits that his fear of speaking in front of groups is preventing him from attaining his consulting objectives, and he is aware that this fear has affected his relationships with relatives, friends, and women. He can see how this hang-up affects his life. For example, if he is asked, "What do you think the group might think if you speak poorly to them?" He might answer, "Oh, I have this feeling every once in a while that I'm stupid and I don't want others to find out that I'm not as sharp as I appear." When asked if he would consider receiving some personal counselling that would handle that fear he would respond, "Sure, I'll do anything it takes. I want to get on with life and live it—I'm tired of hiding. Who can help me with this?"

After some professional counselling this person will move on to reach his objectives. He knew all along that the major barriers to overcome were internal and personal. The successful personality (1) recognizes and identifies his personal and private fears, (2) is willing to communicate intimately about them, and (3) looks at them as barriers to his goals and is willing to get assistance to knock them out of the way.

Contrast this with the unsuccessful person. When asked what fears he has about going off on his own he'll say something like, "I have a lot of family responsibilities and I can't take any chances. Maybe I'll try consulting when I'm ready to retire." He never even answered the question. When asked about personal hang-ups that have prevented him from reaching his goals, he might say something like, "I've got it pretty much together. No big problems in life. My job is a little boring and my relationship with my wife isn't what it used to be, but I guess I'm doing all right. The reason I haven't started my own business is that credit is tight and I can't get the required start-up capital." This type of person thinks there is something wrong with having or admitting fears or hang-ups. His failures and lack of progress are always due to some external reason; they are never related to himself.

We all have fears when embarking on a new adventure. Either we overcome the fears and lead an exciting life, or we let our fears overcome us and we lead a life of mediocrity. Since 1974 I've investigated many methods which help successful professionals overcome their personal barriers. The most successful methods that I have found have been developed by L. Ron Hubbard. His book, *Dianetics, the Modern Science of Mental Health* is an excellent reference for the entrepreneur who is interested in personal development. This book can be obtained from Bridge Publications, 4833 Fountain Avenue, East Annex, Los Angeles, California 90029.

My Personal Philosophy

My personal philosophy in life is "Find something that you enjoy and then get damn good at it." Almost any pursuit in life can become a richly rewarding experience if it is taken seriously and professionally, and technical consulting is certainly no exception. I have found no greater thrill in life than overcoming personal barriers, developing excellence at my trade, and leaving in my wake many satisfied managers. You can do it too. Wishing you success!

Appendices: Sample Contracts

I am including the following section on contracts because it gives a broad perspective on the different types of contracts with which a consultant should be familiar. The introduction and contracts for appendices A–J were provided by the Industrial Designers Society of America, 6802 Poplar Pl. #303, McLean, VA 22101. The contracts for appendices K and L are contracts that I personally use in my business. The contract for appendix M was provided by the National Society of Professional Engineers, 2029 "K" St. N.W., Washington, D.C. 20006. The contract for appendix N was provided by The Hughes Aircraft Company, P.O. Box 90515, Los Angeles, CA 90009. I would like to gratefully acknowledge IDSA, NSPE, and Hughes Aircraft for their kind support.

No claim is made for the complete legal accuracy of the sample contracts that are provided in this appendix. These contracts are provided so that you can get an idea of how contracts are written. When you enter into a contractual agreement with your client, the validity of that agreement depends on the understanding that you have mutually generated before signing the contract. If you have any doubts about the validity of your contract, you should contact an attorney who can legally represent you in the proper manner.

INTRODUCTION

Contracts between industrial designers and their clients have been changing over the years. In the early years of the profession most of the contracts were based on those in use by other professions and on standard legal practice.

There were several reasons for this: First, there was no precedent on which to base a contract form which would fit the special needs of the industrial designer/client relationship. Second, early designers were doing everything possible to establish industrial design as a profession; a contract written in legal terminology contributed to the impression of a solid, stable profession. Third, such contracts suggested that industrial designers operated on a sound business-like basis.

As the years have gone by and the profession has, in fact, become established, contract forms have been developed which are adaptable to almost any designer/client need. Designers have been as creative in developing contract forms and types to fit the needs of their potential clients as they have been creative in their design efforts. Today, contracts fall into two basic categories:

In the first category, the keynote is simplicity—the contract is usually a short informal letter. When the work to be performed is easy to define, and the scope of services is simple, this kind of brief document is usually sufficient to cover the needs of both client and designer. This is particularly true when a friendly designer/client relationship has been established over a period of time. The client has, of course, become familiar with what he can expect from the designer, and the designer wants to simplify his paperwork. Much of the work performed today on a project basis is covered by a letter-of-agreement, often only one page in length. Letters-of-agreement are in wide use, even when the project is of great importance and substantial fees are involved. Naturally, formal legal language seldom appears in this contractual form.

In the second category are longer and more formal contracts developed to cover complicated programs. As designers offer a wider and wider range of services, in order to solve the complex problems submitted to them by their clients, more complicated contracts are being written. Since the work to be performed is more difficult to define, or perhaps just more involved, great care is being exercised in the terminology used, in definitions, in phrasing, and in specifying as carefully as possible what the designer is going to do and what benefits the client can reasonably expect from these services. However, even these more complex agreements generally use less formal legal language than contracts in the past. Contracts written in formal legal language today are primarily required only for work with very large companies.

The form of complex contracts is rarely a letter-of-agreement, for the provisions are too detailed for informal treatment. A legal contract form is sometimes used, but more often a new form of contract, called the Proposal/Contract, is used. This new contract form has come into very wide use in the last 10 to 15 years. It performs the dual function of serving both as a proposal and as a contract. The fundamental difference between the Proposal/Contract and other forms of contracts is that it acts as a selling document as well as a contract. It states the background reasoning for the program; defines the objectives of the program and the benefits the client may reasonably expect to obtain; outlines in detail the specific procedures which will be used; adds the standard factors normally covered by all forms of contracts—time schedules, estimated costs, general conditions governing client/designer relationships, and finally provides space for formal acceptance of the program by the client.

The Proposal/Contract is more flexible and more complete—allowing a wider variety of approaches—than a standard legal contract form. It is a direct, self-sufficient document, outlining the designer's ideas to others, without third-party interpretation, and thus allows the executive with whom the designer is negotiating to pass the document on to others in the company who are not in direct personal contact with the designer. It has developed over the years into the form of contract which most readily fits the varying needs of the designer; it is, also, most closely attuned to corporate procedures and requirements.

GENERAL GUIDELINES

A number of fundamental factors concerning contracts should be kept in mind:

First, clarification is of the greatest importance, and clarification usually means simple, precise terminology and brevity. However, make certain that essential factors have been covered completely.

Second, no two contracts will be exactly alike. The form of the contract, and the type of arrangement should be directly related to the needs of the client and the work to be performed.

Third, be careful in using "designer's language" unless it is certain that the client is aware of the exact meaning of these words and phrases, and appropriate definitions have been incorporated into the contract.

Fourth, the development of an appropriate contract is definitely a part of the creative process of design. To a great extent, the actual design process which follows is based on the creative thinking and planning built into the contract.

Fifth, never work without a written and signed agreement. Both the client and the designer are subject to loss of memory, misunderstanding, and misinterpretation when a verbal agreement is not committed to writing. Experienced designers insist on discussing all the provisions of a client/designer relationship in detail, so that there is very little possibility for misunderstandings later. This is usually effective in convincing the client that a contract is needed. Should a client resist signing a contract, you can be reasonably sure that there is something questionable about the project. Conversely, should a client be readily agreeable to signing a contract, you can be reasonably sure that the project is sound.

Sixth, it is definitely advantageous for the designer to write the contract and submit it to the client, rather than the reverse. Even if there are changes and additions, most of the provisions will remain. When a client prepares a contract, his legal advisors may include restrictive clauses which are, of course, most protective of the client. When the designer feels he must challenge such provisions, he often has difficulty averting a feeling of suspicion on the part of the client. It is easier to challenge a new provision which the client requests be added, than to ask that a provision in a contract prepared by the client be deleted.

Seventh, it is advisable to obtain professional legal advice concerning any document a designer may sign.

Appendix A: Designer-Prepared Contract

Form: Letter-of-Agreement

Type: Fixed-Fee

Client Action Required: Issue a Letter of Authorization, or a Purchase Order

INDUSTRIAL DESIGN SERVICE ASSOCIATES, INC.

200 Middle Road   Chicago, Illinois

American Manufacturing Corporation   January 2, 1970
5000 Fifth Avenue
New York, New York

Attention: Mr. James Jones

Gentlemen:

We appreciate very much the opportunity to submit this proposal for the design of a new plastic milk case to hold a minimum of sixteen quarts or eight half gallons of milk.

It is our understanding that this milk case will have the following features. It is to be:

1. Stackable.
2. Light in weight.
3. Of at least the same strength as the present wooden cases.
4. Distinctive in appearance.
5. Durable as the present case (seven years).
6. Designed to be used in the new vertical dairy cases in super markets.

The new milk case is also to attempt to solve the problem of inventory (first in, first out), the problem of stacking (at least five high), and to keep in mind its possible use also in the soft drink field.

In order to accomplish the solution to these problems, our work will evolve in three steps:

1. DESIGN RESEARCH

    We will need to acquaint ourselves fully with all the research which American Manufacturing Corporation has already done in the solution of this problem. We will need to know intimately, the molding and manufacturing problems. We will visit with your customers to observe their problems in delivery, filling of cases, and return of cases to the plant.

2. DESIGN

    After reviewing all possible research we will then start design studies, keeping in mind the basic points listed above.

3. PRODUCTION—FOLLOW THROUGH

    After acceptance of our design suggestions by American Manufacturing Corporation, we will be available for consultation work in the actual production of these cases in order to assure American Manufacturing Corporation, and ourselves, that the newly designed case meets all criteria.

COMPENSATION

As compensation for this assignment we will require a budget for design time of approximately $16,000. All time will be billed to you on a man-hour basis as the time accrues. We expect the duration of this assignment to be six months.

EXPENSES

Industrial Design Service Associates, Inc., is also to be reimbursed for all purchases of photostats, blueprints, typography, photography, models, etc., as requested and approved by you at cost, plus a ten percent handling charge.

Travel, when required and approved by you, will be charged at actual cost.

PATENT ASSIGNMENTS

We agree to communicate to you any and all inventions and improvements conceived or made by us or any of our employees prior to termination of this agreement relating to the subject matter of the project, whether made solely by us or jointly with you, and all such inventions and improvements shall become your sole and exclusive property without any obligation to make any payment to us or our employees therefor except such sums as you may be obligated to pay pursuant to this agreement between us. We agree to execute and to cause our employees to execute patent or copyright applications, assignments thereof to you.

We will be available for further discussions of this proposal at any time. The terms of this proposal will be held for thirty days from January 2, 1970.

        Sincerely,

        INDUSTRIAL DESIGN SERVICE ASSOCIATES, INC.

        Ronald Smith
        Vice President

RS:rl

Appendix B: Designer-Prepared Contract

Form: Legal Contract

Type: Retainer: Minimum Dollar Volume Guarantee

Client Action Required: Sign and Return One Copy to Designer

AGREEMENT

BY THIS AGREEMENT, made this 2nd day of January, 1970, the AMERICAN MANUFACTURING CORPORATION, 5000 Fifth Avenue, New York, New York (herein after referred to as "AMCORP"), a Delaware corporation, and INDUSTRIAL DESIGN SERVICE ASSOCIATES, INCORPORATED, 200 Middle Road, Chicago, Illinois (herein after referred to as "IDSA"), an Illinois corporation, mutually agree with each other that:

1. IDSA shall be engaged by AMCORP as its sole industrial design consultant in the area of major electric appliances—consisting specifically of refrigerators and freezers, ranges, dishwashers, room air conditioners, and laundry equipment. As such industrial design consultant, IDSA shall act as an independent contractor working in conjunction with representatives of AMCORP.
2. As design consultant, IDSA shall make the following services available to AMCORP:
    a. Ronald Smith, individually, and such other officers and personnel of IDSA as are appropriate, shall be available to AMCORP for counsel, advice, and guidance on any matters involving design in the area of major appliances.
    b. Thomas Williams, IDSA Vice President, will be assigned as account executive. His responsibilities will be:
       (1) To act as coordinator between AMCORP and IDSA
       (2) To properly interpret AMCORP needs to IDSA.
       (3) To administrate AMCORP work appropriately within IDSA to meet these needs.
    c. IDSA will be available to the full extent desired by AMCORP, in accordance with a program and schedule to be jointly developed by AMCORP and IDSA, for the performance of the following additional types of design services:
       (1) Development of the design themes, or patterns, for the above mentioned major product lines.
       (2) Origination and development of new product concepts or new product features.
    d. Appropriate IDSA personnel will provide advice and counsel, as required, to the AMCORP engineering and design groups during their development of final production details of the complete product lines, for which IDSA has previously developed the design themes.
3. The services described above will be performed under the following general procedure:
    a. Initially, a series of meetings will be scheduled between appropriate AMCORP and IDSA personnel, to develop, jointly, an over-all program outline for the 6- to 9-month period ahead.

    b. Following this, meetings will be scheduled—approximately once each month, or in accordance with current needs—between appropriate AMCORP and IDSA personnel:
        (1) For review of progress and modification of the over-all program.
        (2) To maintain a high degree of coordination between the activities of various AMCORP groups and IDSA.
    c. Projects for the development of design themes for the various product lines will be initiated in accordance with the program previously developed. The scope and specifics of each individual project will be determined jointly by AMCORP and IDSA, and IDSA will then submit a budget and time estimate to AMCORP, for approval, before work is begun. Such individual projects may run in sequence, or concurrently according to current needs.
    d. Specific and detailed recommendations, in writing, complete with budget and time estimates, will be submitted to AMCORP by IDSA, on its own responsibility and according to its own judgement, for the initiation of projects for the development of new product concepts or new product features. AMCORP approval will be necessary before work is begun.

4. As design consultant, IDSA agrees to be available for the performance of the foregoing services to the full extent desired by AMCORP.

5. For the program outlined above, IDSA shall be paid a guaranteed minimum of One Hundred Eighty Thousand ($180,000.00) Dollars per year, covering the cost of all IDSA personnel services. Invoices will be submitted for payment at the end of each month, covering the actual effort expended during that period.

In addition, IDSA shall be reimbursed by AMCORP for the actual costs of travel, outside the Chicago area, by IDSA pesonnel, as required in the execution of the above described program. Invoices will be submitted monthly for travel reimbursement.

In addition, IDSA shall be reimbursed by AMCORP for the actual costs, plus a twelve (12%) per cent handling charge, of all out-of-pocket costs, including blueprints, photostats, photoclearlines, typesetting, special modelmaking materials, and photographic supplies. Invoices will be submitted monthly for out-of-pocket cost reimbursement.

It is understood that travel and out-of-pocket expenses do not contribute to the fulfillment of the $180,000.00 committment.

The scope of work for each project handled under this program will be defined in writing by AMCORP. Based on these written definitions, IDSA will then submit a not-to-exceed cost estimate for each project. Work will not begin on a project until IDSA has received authorization from AMCORP to proceed.

APPENDIX B

Working procedures, progress of the work, and the status of the yearly budget will be reviewed quarterly by AMCORP and IDSA principals.

In the event that the volume of work requested and authorized by AMCORP does not total $180,000.00 in IDSA personnel services during the contract year, the balance—the difference between the invoiced total and $180,000.00—will be due and payable to IDSA on the last day of the contract year.

In the event that the volume of work requested by AMCORP during the contract year exceeds a total of $180,000.00, the IDSA coordinator is charged with the responsibility of informing AMCORP of this fact before work exceeds the guaranteed minimum budget.

6. IDSA will endeavor, in good faith, to keep confidential all information regarding the activities of AMCORP and shall obligate all its employees to do likewise.

7. IDSA will not render services similar to those described above for any competitor of AMCORP in the specific areas of major electric appliances described in paragraph 1, above, during the term of this agreement.

8. IDSA agrees that the services hereunder shall be for the sole and exclusive benefit of AMCORP and that it will, when requested by AMCORP, divulge and disclose to it, its officers, employees, and representatives, the progress of the services hereunder and all designs, drawings, sketches, mock-ups, models, ideas, inventions, and other results evolved or conceived in connection with service under this agreement and it further agrees that all designs, drawings, sketches, mock-ups, models, and samples made under this agreement shall be the property of AMCORP and shall be turned over to AMCORP upon request of AMCORP.

9. IDSA will promptly disclose and assign to AMCORP all inventions, including mechanical and design inventions, made or created, during the life of this agreement, by IDSA, its officers, associates, and employees relating or pertaining to any product or products upon which IDSA has undertaken services hereunder, together with such patent or patents as may be obtained thereon, in this and all foreign countries and will execute or have executed all proper papers for use in applying for, obtaining, and maintaining such United States or foreign patents thereon as AMCORP may desire and will execute or have executed and deliver or have delivered all proper assignments thereof, when so requested, but at the expense of AMCORP.

IDSA represents that the conditions of employment or the terms of the contract of employment of any person engaged in service under this agreement shall include provisions which will secure to IDSA the right to disclose and assign to AMCORP the inventions specified in this paragraph.

10. This agreement shall be effective January 1, 1970, and may be terminated by either party on December 31, 1970, and thereafter on any anniversary of that date upon three (3) months' prior written notice, whereupon neither party shall have any other or further rights against the other except with respect to rights which accrue prior to the effective date of such cancellation.

AMERICAN MANUFACTURING CORPORATION

By _____

INDUSTRIAL DESIGN SERVICE ASSOCIATES, INC.

By _____
    Ronald Smith, President

Appendix C: Designer-Prepared Contract

Form: Legal Contract

Type: Retainer: Time-and-Materials

Client Action Required: Sign and Return One Copy to Designer

AGREEMENT

BY THIS AGREEMENT, made this 2nd day of January, 1970, the AMERICAN MANUFACTURING CORPORATION, 5000 Fifth Avenue, New York, New York (herein after referred to as "AMCORP"), a Delaware corporation, and INDUSTRIAL DESIGN SERVICE ASSOCIATES, INCORPORATED, 200 Middle Road, Chicago, Illinois (herein after referred to as "IDSA"), an Illinois corporation, mutually agree with each other that:

1. IDSA shall undertake such design and development programs for the development of new and/or better AMCORP products as are individually assigned by AMCORP on written purchase orders directed to IDSA by AMCORP'S Purchasing Division and accepted by IDSA.

2. The parties recognize that IDSA may perform services for AMCORP on programs not governed by this agreement, and that instances may arise in which purchase orders drawn under this agreement are not acceptable to IDSA. In any such case where IDSA so performs services for AMCORP during the term of this agreement, the parties shall provide therefor by separate agreement.

3. AMCORP shall pay IDSA for the services to be performed by IDSA on a time-and-material basis.

    "Time" will be charged to AMCORP on the basis of the actual invoicing rate applicable to each employee within the class of employees and rates set forth in the purchase order.

    "Material" will be charged on an actual cost basis.

    Travel Expense, when specifically authorized by AMCORP, shall be charged to AMCORP at actual cost.

    All records supporting charges made by IDSA shall be subject to audit by AMCORP upon ten days notice.

    In no event, however, shall AMCORP be charged, or liable to pay, in excess of an amount, to be set forth in the purchase order, as a "not to exceed" amount for the performance of the design and development program.

4. IDSA shall issue invoices each month for services rendered during the preceding month. Such invoices shall be payable net thirty days.

5. IDSA shall not, during the term of an assigned and accepted project, perform services similar to those covered thereunder for any other person, firm, or corporation developing, manufacturing, or promoting the sale or use of materials similar to those of AMCORP. IDSA shall not at any time, either during or subsequent to the performance of this agreement, directly or indirectly disclose to others any information received from AMCORP concerning the designs, processes for the manufacture, or technical data with respect to any such concept or project, except as is necessary for the development of the concept or to conduct the project, without the express written consent of AMCORP.

6. If any officer or employee of IDSA shall make any invention involving the subject matter hereof, during or as a result of the performance of the services covered by this agreement, such invention shall be disclosed promptly to AMCORP and shall be the property of AMCORP. IDSA shall procure the execution of any application for patent thereon which AMCORP may see fit to have prepared, together with an assignment to AMCORP of the entire interest therein for the United States and all foreign countries. All legal or other expenses connected with the execution, registration, or assignment of such patents shall, however, be borne by AMCORP. IDSA represents that it has written agreements with all its officers and employees to be assigned to the work called for hereby, binding them to perform as stated herein.

7. AMCORP shall own all right, title, and interest, including all rights at common law, all copyrights, and all rights to copyright renewal, in and to all material originated pursuant to this agreement, including material furnished to IDSA by third parties. In the case of such material originated by third parties, IDSA shall procure the assignment of all right, title, and interest as aforesaid to AMCORP. "Material" as used herein, includes, without limitation, verbal compositions, drawings, photographs, motion pictures, designs, music, three-dimensional figures, layouts, diagrams, tables, surveys, reports, merchandising plans, ideas, slogans, characters, trade-marks, and trade names.

8. IDSA hereby represents, warrants and guarantees that each design concept and the description, drawing, model or other product resulting from any design project furnished by it pursuant to this agreement, shall be an original conception, that is, not copies from any prior work, of IDSA; provided, however, that this is not to be construed as a warranty or guarantee that there may not have been prior similar conceptions in the art nor is it to be construed herefrom that IDSA assumes the obligation for searching the art for any such prior similar conceptions either in the field of publications, use, or letters patent in any country.

9. If, in the presentation or merchandising of design concepts to its customers, AMCORP shall require additional services from IDSA, including but not limited to the cost of displays, transportation of materials, and the services of IDSA personnel not otherwise covered hereunder, IDSA shall not contract therefor on behalf of AMCORP without written authorization. In any event, IDSA, if ordered to provide such additional services by AMCORP, shall be reimbursed for costs incurred thereby and compensated for services of its personnel, but such compensation shall not include any charge for services of any of the officers of IDSA. IDSA shall submit detailed invoices covering these costs, which invoices shall be subject to audit by AMCORP.

10. This agreement shall be effective as of the day and year first above

written and shall terminate one (1) year thereafter. Articles five (5), six (6), seven (7), and eight (8) of this agreement shall continue in force after termination.

IN WITNESS WHEREOF, the parties hereto have caused this to be executed by their duly authorized representatives the day and year first above written.

AMERICAN MANUFACTURING CORPORATION

By _____
    General Manager-Purchasing

INDUSTRIAL DESIGN SERVICES ASSOCIATES, INC.

By _____
    President

Appendix D: Designer-Prepared Contract

Form: Legal Contract

Type: Retainer for Consultation

Client Action Required: Sign and Return Duplicate

AGREEMENT OF JANUARY 1, 1970
Between
AMERICAN MANUFACTURING CORPORATION
(hereinafter called AMCORP)
And
INDUSTRIAL DESIGN SERVICE ASSOCIATIONS, INC.
(hereinafter called IDSA)

COVERING INDUSTRIAL DESIGN CONSULTATION SERVICES ON OFFICE FURNITURE AND ASSOCIATED EQUIPMENT.

I. IDSA will perform the following services for AMCORP:
   A. Study the design problem related to:
      The present product line.
      Competitive products.
      Production possibilities and limitations.
      Operational and human engineering requirements.
   B. Consult in AMCORP's offices to:
      Develop preliminary sketch layouts.
      Check layout and detail drawings in work.
      Advise on the development of necessary prototype and/or pre-production model drawings.
      Make specific recommendations as to materials, colors, finishes, human engineering factors, etc.
      Review production drawings and final models made by AMCORP.
   C. Obtain specialized services (modelmakers, renderers, graphic artists, etc.) as required and as approved in advance by AMCORP.
   D. Co-operate closely with AMCORP officials and engineering executives to insure the proper application of design recommendations up to the point of production release.

II. AMCORP will, in return for the above services, pay IDSA:
   A. Consultation Retainer Fee: Twelve Thousand Dollars ($12,000) payable at the rate of One Thousand Dollars ($1,000) per month for twelve (12) consecutive months, the first payment being due on the date of this Agreement. This fee covers:
      1. Exclusive consultation services of office furniture and associated equipment.
      2. Twelve (12) days consultation in AMCORP's offices. (See Note 1.)
      3. Development time in IDSA's offices for four (4) hours per month. (See Note 2.)
         NOTE 1: Additional in-plant consultation will be made available at the rate of Three Hundred Dollars

($300) per day, including travel time, if requested and approved in advance by AMCORP.

NOTE 2: Additional development time in this office will be made available at the rate of Thirty Dollars ($30) per hour based on quarter-hour increments. Any time required beyond ten (10) hours per month total to be approved in advance by AMCORP.

B. Outside services (modelmakers, draftsmen, renderers, graphic artists, etc.) at cost, as approved in advance by AMCORP.

C. Direct charges (telephone, blueprints, photographs, reproduction art work, etc.) at cost.

D. Travel and subsistence at cost.

NOTE: Items B through D to be billed at the end of the month in which incurred.

III. Any patents resulting from the work of IDSA will be assigned to AMCORP. Inventorship to be determined by AMCORP's patent counsel. Patent costs to be paid by AMCORP.

IV. During the term of this Agreement IDSA will not design or advise concerning any product directly competitive to those described in the heading of this agreement.

V. The above services shall continue for a period of Twelve (12) consecutive months from the date of this Agreement, and succeeding twelve-month periods thereafter unless terminated by either party on Sixty (60) days notice delivered in writing by registered mail.

ACCEPTED BY:

AMERICAN MANUFACTURING CORPORATION	INDUSTRIAL DESIGN SERVICE ASSOCIATES, INC.
By _____	By _____

Appendix E: Designer-Prepared Contract

Form: Proposal/Contract

Type: Project: Cost-Plus-Fixed-Fee

Client Action Required: Sign and Return Duplicate

PROPOSAL
FOR THE DEVELOPMENT OF
ELECTRIC POWER OVERHEAD TRANSMISSION SYSTEMS AND
COMPONENTS
FOR THE
AMERICAN MANUFACTURING CORPORATION
By
Industrial Design Service Associates, Inc.
January 2, 1970

I. INTRODUCTION

Relative to our recent discussions, we would like to propose, for your consideration, a project for the development of new *Electric Power Overhead Transmission Systems and Components*.

II. BACKGROUND

In the recent past, particularly the last decade, the use of electrical power has grown at a fantastic rate. In the period of the next ten years, it is estimated that the need for electrical power in the U. S. will at least double. This phenomenal growth is primarily caused by the ever-growing acceptance of electricity as a safe and convenient source of power by the consumer, by the decreasing cost of such power, and by the population explosion.

Practically every power company in the U. S. finds itself critically short in its power-generating facilities; and, in addition, the transmission systems are not only generally outdated, but are frequently inadequate for the new loads needed. New communities have sprung up at a rapid rate, and additional volumes of electricity are being requested by present consumers throughout the entire country. As new communities develop, electrical power is extended to them from systems originally installed to serve much smaller areas and loads. The result is that throughout the U. S. today, our transmission systems are a complex of modifications—usually bearing little resemblance to the original system design—developed to answer the practical needs of the moment.

The visual hodge-podge of overhead electrical systems, which surrounds us today everywhere—forests of transmission poles and towers, overhead wires running in every direction—has led to ever-growing public indignation. Since electrical needs are increasing, this can only result in additional poles and wires in the future, and probably increasing public resistance to such confusion and ugliness.

There is an excellent opportunity for major suppliers to the electrical industry to develop new concepts in the entire area of overhead power distribution which will contribute to solutions of these fundamental problems. If new concepts can be developed which provide increased acceptance by consumers—improved functional and visual organization of overhead power lines, improved functional and visual designs for transmission poles and structures, and improved functional and visual designs for substations—and if these new concepts can be manufactured and installed at costs comparable to presently used system components, power companies should welcome such developments with high enthusiasm. In fact, even if new proposed systems were of somewhat higher cost, if they presented solutions to these basic problems, they would undoubtedly receive favorable acceptance.

For those major suppliers, such as AMCORP, whose products lend themselves primarily only to overhead transmission systems, a move to underground systems would be of critical importance.

Substation design is today completely outmoded in architectural appearance. Distribution substations, in their wide variety of form, are particularly incompatible with their surroundings. Often they appear as industrial buildings, switch- and transformer-yards located in the midst of choice residential areas. Public reaction to such installations located close to their homes is growing daily. The entire area of substations and their equipment is a fertile area for both functional and design improvements.

The design of equipment components—pole-mounted transformers, for example, which appear on practically every street corner in the U. S.—has received very little attention in recent years. Developments have proceeded in the direction of improved function, and cost saving, but little or no attention has been given to their compatibility with the background in which they function. The entire area of equipment is open for the development of new concepts.

Fundamentally, however, rather than a piecemeal approach to the subject, the opportunity seems ripe, due to the magnitude of the problem, for AMCORP to develop and present new concepts for the *entire* overhead transmission system.

Up to now, AMCORP developments for the electric power industry have concerned themselves primarily with specific components. The reaction to and acceptance of these AMCORP developments, both by the power companies and by manufacturers of electrical equipment, have been excellent. These reactions indicate clearly that an opportunity exists for even greater strides forward. By approaching the problem on a broader level, the top managements of the larger power companies can be approached. Decisions of such magnitude will lie with this group, rather than with the middle echelons of management and the systems engineers who have previously been approached.

By developing complete new system concepts which assist in resolving the basic problems mentioned above, and by utilizing AMCORP components logically in these new systems, presentations can be made to the 25 or so "decision-making" power companies in the industry—presentations which can substantially improve the acceptance and use of AMCORP products by the industry.

III. OBJECTIVES

To develop a group of new system concepts for the overhead distribution of electric power from cross-country transmission systems to individual points of attachment in the home. This will include consideration of substations, transmission structures, and system components.

To improve the acceptance and use of AMCORP components by the electric power industry by demonstrating, through these new system concepts, the practical advantages and tangible benefits of AMCORP products.

To demonstrate, through these new system concepts, that AMCORP can materially assist the power companies to solve the basic problems described above in "Background."

IV. PROGRAM/PROCEDURE

We suggest that a program be developed in two phases, as follows:

Phase I: Development of New Concepts

1. Further research and study of the practical problems referred to in "Background."
2. Development of a variety of new concepts for overhead transmission systems.
3. Development of preliminary designs for specific typical components of the systems. The specific components to receive design attention will be jointly chosen, following evaluations of the overall system concepts, by AMCORP and IDSA.

The above work will be developed in chart, diagram, and sketch form. Rough scale mock-ups may be developed, where deemed necessary for clarity.

Presentation of the above work to AMCORP for discussion and evaluation. Following this, AMCORP may wish to obtain the objective reactions of a carefully selected group of representatives from power companies, electric equipment manufacturers, and consultant engineering organizations active in this field.

As a result of this evaluation, one or more system designs, and a group of specific components will be chosen for further development.

Based on these decisions, plans will be formulated for presentation of this material to the industry. The form of presentation, and methods to be used will be determined jointly by AMCORP and IDSA.

Phase II: Further Development/Preparation of Presentation Material

Further development of the system concepts, and the specific components. The form of the development—drawings, models, diagrams, etc.—will follow the requirements of the presentation.

Work in Phase II will be carried to the point that the material will be suitable for presentation to AMCORP customers. However, it is not contemplated that any of the concepts developed in this program will be carried to the final stage of development necessary for actual production.

V. ESTIMATED TIME SCHEDULE

Phase I: 4 months.
Phase II: 6 months.

VI. PROGRAM COSTS

We suggest that costs of the program, as described above, be handled on a cost-plus-fixed-fee basis.

Our fee for this program will be Sixteen Thousand Dollars ($16,000), to be paid in ten equal installments of One Thousand Six Hundred Dollars ($1,600) each. One installment of the fee will be due and payable at the beginning of each of the next ten months.

In addition, costs for our time will be invoiced at the end of each month of the basis of work actually performed during that period. Costs for time of our personnel shall be computed on the basis of 1 unit for actual direct labor cost, plus 1½ units (of direct labor cost) for overhead. Detailed invoices will be submitted for payment each month indicating the number of hours each IDSA individual has worked on this program, his direct labor rate, and the total of his costs, based on the formula above.

In addition, travel expenses of IDSA personnel, when involved in work directly related to this program, and all out-of-pocket costs incurred in the execution of this program will be invoiced to AMCORP at the end of each month for reimbursement, at our actual cost.

VII. GENERAL CONDITIONS

A. Inventions

Any and all inventions, improvements, developments, patent applications, and patents relating in any way to the subject matter of this agreement, made by any member or employee of IDSA as a result of this agreement or during the term thereof and one year thereafter, shall be promptly disclosed to American Manufacturing Corporation, shall be assigned and hereby are assigned to American Manufacturing Corporation and shall be American Manufacturing Corporation's exclusive property. IDSA shall make full and prompt disclosure to American Manufacturing Corporation of all such inventions, improvements, and developments and shall procure the execution of all necessary documents and otherwise assist American Manufacturing Corporation in every proper way, so that American Manufacturing Corporation may obtain patent protection thereon and secure priority rights thereto in any country in the world; but all legal or other expenses incurred in connection therewith shall be borne by American Manufacturing Corporation. IDSA represents that it has agreements with all of its members and employees who will perform work called for by this agreement binding them to perform as stated herein.

IDSA shall cause to be kept complete and systematic writings, drawings, and sketches, including notes on all work, descriptions, diagrams, and other data pertaining to the work performed under this agreement; and IDSA shall deliver all such materials to American Manufacturing Corporation upon termination of this agreement.

IDSA recognizes that American Manufacturing Corporation has valuable property rights in its trade secrets, designs, and confidential information. IDSA shall not at any time disclose to any third party or make use of, except for the benefit of American Manufacturing Corporation, any process, device, structure, know-how, technical data, trade secret, machine, or composition of matter or other information acquired from American Manufacturing Corporation or which relates to or results from investigations or activities under this agreement. American Manufacturing shall own all right, title, and interest, including all rights at common law, copyrights, and rights to copyright renewal, in all material originated pursuant to this agreement; and IDSA shall procure the execution of all necessary documents to vest all such right, title and interest in American Manufacturing. Material, as used herein, shall include without being limited to verbal compositions, drawings, photographs, designs, three-dimensional figures, layouts, diagrams, surveys, reports, ideas, slogans, characters, trademarks, and trade names.

The foregoing provisions of this general section shall survive termination of this agreement.

IDSA represents that it does not have at this time and will not have during the term of this agreement and for six months thereafter any conflicting clients, interests, programs, projects, or agreements, without the written consent of American Manufacturing Corporation.

VIII. CONCLUSION

Should AMCORP wish to carry any of the concepts originated in this program to a final stage of development, IDSA will prepare additional project or program proposals to cover all such work.

IX. ACCEPTANCE

If the proposal, as outlined above, is agreeable to you, please be kind enough to sign the copy and return it to us, and retain the original for your own files.

INDUSTRIAL DESIGN SERVICE ASSOCIATES, INC.

Ronald Smith
Vice President

Accepted:

AMERICAN MANUFACTURING CORPORATION

By _____

Title _____

Date _____

Appendix F: Designer-Prepared Contract

Form: Proposal/Contract
Type: Fixed Monthly Charge
Client Action Required: Sign and Return Duplicate

PROPOSAL
FOR THE DEVELOPMENT AND IMPLEMENTATION
OF A
CORPORATE DESIGN PROGRAM
FOR
AMERICAN MANUFACTURING CORPORATION
Industrial Design By Service Associates
January 2, 1970

A. INTRODUCTION

Following your request, we are pleased to submit our proposal for the development and implementation of a corporate design program for American Manufacturing Corporation. The program will be concerned with the products and product lines of the Air Conditioning, Major Appliance, Controls, Machinery, and Industrial Divisions.

B. OBJECTIVES

1. To evaluate the design of the most important products or product lines presently being marketed by each American Manufacturing Corporation Division.
2. To compare the design of these American Manufacturing Corporation products with the design of major competitive products.
3. To present recommendations for specific products or product lines in each Division, which should receive immediate design attention . . . and why.
    Priorities, design objectives and budget estimates, and time schedules will be prepared for each recommendation.
4. To submit recommendations for methods and procedures which will ensure that every product produced by the corporation in the future will receive authoritative design consideration.

C. PROCEDURE

Phase I: Study of American Manufacturing Corporation Products and Product Lines—Study of Comparable Products of Major Competitors

1. Visits will be made to the Air Conditioning, Major Appliance, Controls, Machinery, and Industrial Divisions, to become acquainted with:
    a. Division objectives.
    b. Products and product lines.

c. Current product-development procedures.
d. Markets and marketing practices.
e. Major competitors' products and practices.
2. Following these visits, an independent study will be made, firsthand—to the extent physically possible—of the most important products produced by the major competitors of each American Manufacturing Corporation Division.

Phase II: Formulation of Detailed Design Program for Each Division

1. Based on the investigation and analysis completed in Phase I, a program for specific design action will be detailed which will:
   a. Specify the products or product lines of each Division which should receive immediate design attention, and the reasons why.
   b. Provide a list of objectives for redesign of each of these products or product lines.
   c. Present priority, timing, and budget estimates for each design project.
   d. If appropriate, present suggestions for new products.
   e. Recommend the method and procedure by which design should be handled in each Division in the future.

Phase III: Implementation of Design Projects

1. Based on the recommendations provided in Phase II, American Manufacturing Corporation may, at its discretion, authorize IDSA to proceed with specific design projects.
2. In the event that, prior to the completion of this program, American Manufacturing Corporation desires to implement one or more specific design projects, American Manufacturing Corporation may authorize IDSA to proceed with such work concurrently, under separate project agreements.

D. COST AND DURATION

We estimate that the study and recommendation program, as outlined above, will require eight months to complete.

Our fee for the program will be $6,000 per month. Invoices will be submitted for payment each month, in this amount. Expenses for travel will be billed, at our cost, in addition.

E. CONTRACTS FOR RESPONSIBILITY

Mr. Ronald Smith will be directly in charge of the work, assisted by such other Partners and staff members of IDSA as are appropriate.

F. GENERAL

Our work for clients is conducted on a confidential basis, and we shall treat all information developed hereunder in accordance with established professional standards.

All material developed under this program shall become the sole property of American Manufacturing Corporation.

G. ACCEPTANCE

We have prepared this proposal in duplicate, so that, if acceptable, you may sign one copy and return it to us, and retain the other for your files.

INDUSTRIAL DESIGN SERVICE ASSOCIATES

Ronald Smith, Managing Partner

Date _____

Accepted:

AMERICAN MANUFACTURING CORPORATION

By _____

Date _____

Appendix G: Designer-Prepared Contract

Form: Royalty Agreement

Client Action Required: Sign Copy and Return to Designer

LETTER OF AGREEMENT

American Manufacturing Corporation   January 2, 1970
5000 Fifth Avenue
New York, New York

Gentlemen:

It is proposed that the American Manufacturing Corporation (AMCORP) retain the services of Industrial Design Service Associates, Inc. (IDSA) on the following terms.

1. IDSA shall, during the term hereof, act as design consultants to AMCORP and will render all services, as required by AMCORP from time to time, in connection with the development of designs of new products and the redesign of existing products, including, without limiting the generality of the foregoing, the determination of external esthetic design characteristics, color combinations, design research, development and construction of design prototypes, drawings, and specifications relating thereto.

2. In view of the confidential nature of the services being and to be rendered and the fact that AMCORP will make disclosures to IDSA of matters which constitute AMCORP trade secrets or information not generally known to the trade, IDSA, as an express condition hereof, agrees:

    (a) During the term of this arrangement, it will not, directly or indirectly, perform or render similar services for any other manufacturer or seller of equipment, directly competitive with AMCORP unless AMCORP gives prior consent thereto.

    (b) Neither it nor its employees will disclose, publish, or make known to or for the benefit of any person, other than AMCORP, any information or knowledge not then in the public domain relating to AMCORP products, designs and services being rendered for AMCORP, or any other information not then publicly known in connection therewith through public channels, except as such disclosures may be required by law or judicial process, unless AMCORP gives prior written consent to any such use or disclosures. This provision shall survive the expiration of this agreement.

    (c) IDSA shall promptly disclose to AMCORP all information as and when developed in connection herewith and shall permit AMCORP, through its duly designated employees or agent, to examine such information, obtain data known to or in the possession of IDSA relating thereto and for such purpose, in its premises during reasonable business hours. IDSA hereby acknowledges that all designs, ideas, and products heretofore, or during the term hereof, developed in connection with its services for AMCORP is and shall be the exclusive property of AMCORP. When and as requested so to do by AMCORP, IDSA shall execute or cause to be executed all documents pertinent to the application for patent thereof, foreign or domestic, the assignment thereof to AMCORP and will aid and assist AMCORP in the prosecution thereof, it being understood that the costs and expenses for patent search, application, assignment, and prosecution shall be borne by AMCORP.

3. IDSA shall receive the following compensation for its services hereunder, rendered or to be rendered, which amounts shall, unless otherwise specified, be inclusive of all salaries and overhead and other expenses directly related to such services:

   (a) IDSA shall be reimbursed for all time spent by their principals and staff on product development and product design work at actual salary rates plus 50% for overhead. Such reimbursement shall be billed and payable monthly.

   (b) Such additional compensation as may be mutually agreed upon for services in addition to and not related to product design and product development work, i.e., the design of stationery, printed matter, packaging, displays, etc., payable on normal terms as billed.

   (c) Reimbursement for actual expenses incurred in connection herewith for travel, long-distance telephone tolls and authorized purchases of models, samples, typography, reproduction work and other special work, as may be specifically authorized by AMCORP, prior to the incurring of such expense.

4. In addition to the compensation and payments specified in paragraph 3 above, IDSA shall receive royalties computed on the following basis:

   (a) On all AMCORP new products initially designed and developed by IDSA, during the term hereof, a royalty of one-half of one percent of the net sales thereof by AMCORP during the term hereof and for a period of two years following the termination of this Agreement.

   (b) On all AMCORP products which IDSA redesigns (being those AMCORP products not initially designed by IDSA, which heretofore or during the term hereof have been or will be sold by AMCORP and which AMCORP shall request IDSA to redesign, and which shall be thereafter sold in the redesigned form) a royalty of one-quarter of one percent of the net sales thereof by AMCORP during the term hereof, and for a period of two years following the termination of this agreement.

   (c) For the purposes hereof, royalties shall accrue on net sales made commencing January 1, 1970, and until the expiration of two years after the termination of this Agreement. Royalties shall be computed on a calendar quarterly basis and shall be payable on or before 45 days following the conclusion of the preceding quarterly period. AMCORP shall deliver to IDSA, concurrent with each royalty payment, a statement of its net sales of royalty-bearing products by descriptive categories during the applicable period for which such payment is made, IDSA shall have the right, through its designated representatives to inspect AMCORP's books and records relating to its net sales of royalty-bearing products for the period covered by each statement provided (i) such examination is made during regular business hours; and (ii) IDSA makes written request for such examination within 15 days following receipt of each statement. The statement furnished by AMCORP for each period shall be deemed correct and accepted by IDSA unless, within 30 days following receipt thereof, IDSA shall notify AMCORP, in writing, of any objections thereto, specifying the nature and detail of each such objection.

    (d) For the purpose of computation of royalties "net sales" is defined to mean the net amount invoiced by AMCORP to its customer in connection with the sale of the royalty-bearing product, exclusive of freight, service charges, sales, use, or excise taxes (whether added to or included in the invoice price), and less all discounts, allowances, credits, returns, and adjustments allowed by AMCORP.

5. This Agreement shall be deemed to have commenced on January 1, 1970, and shall continue in force to December 31, 1971, and shall thereafter continue from year to year thereafter subject to the right of either party to cancel and terminate this Agreement at any time after December 31, 1971, upon giving at least 90 days' prior written notice of such intention to the other party.

6. Whenever written notices are required, they shall be deemed sufficiently given if addressed to such party at its then principal office and if given, by telegram or first class certified mail, postage prepaid.

If the foregoing correctly reflects the Agreement between us, please execute and return the duplicate hereof.

Very truly yours,

INDUSTRIAL DESIGNERS ASSOCIATES, INC.

By: _____

We hereby accept and approve the above and foregoing Agreement.

Dated: _____

AMERICAN MANUFACTURING CORPORATION

By: _____

Appendix H: Client-Prepared Contract

Form: Legal Contract
Type: Fixed Monthly Charge
Designer Action Required: Sign and Return One Copy to Client

AGREEMENT

This Agreement, made this second day of January, 1970, by and between American Manufacturing Corporation, a corporation organized under the laws of the State of New York, having its principal place of business at 5000 Fifth Avenue, herein after called the company, and Industrial Design Service Associates Inc., 200 Middle Road, Chicago, Illinois, herein after called the designer, witnesseth that:

1. The company retains and contracts with the designer, for a period of ninety days, commencing on the tenth day of January, 1970, to conduct for the company a program of consultation and design, at the rate of $1500 per month, payable to the designer at the end of each month, and the designer hereby accepts such retainer and contract and agrees that such monthly rate of compensation shall be in full for all services rendered by the designer to the company.
2. The above stated program will be for the purpose of designing a new hair dryer based on the mechanical components of AMCORP's existing model XP-300.
3. The designer shall meet and consult with such personnel of the company, as the company shall herein after determine, at such times and places as shall be mutually agreed upon by the company and the designer. The designer shall furnish to the company such color sketches, working drawings, models, color compositions, and black and white mechanicals as the company shall from time to time request.
4. The company agrees to reimburse the designer for all necessary expenses incurred in furtherance of the designer's services for the company such as, but expressly not limited to, black line prints, blueprints, photostats, photography, and typography.
5. Either party may terminate this contract at any time by giving the other party notice in writing of such intended termination at least thirty days before the date upon which such termination is to occur.

In witness whereof, the aforesaid parties have hereunto affixed their hands and seals the day and year first above written.

APPENDIX H

AMERICAN MANUFACTURING CORPORATION

Corporate Seal   BY _____
                   James Jones
                   Vice President, Product Planning

AGREED TO BY:

IDSA

BY: _____
TITLE: _____
DATE: _____

Appendix I: Client-Prepared Contract

Form: Letter of Agreement
Type: Fixed Fee
Designer Action Required: Sign and Return Duplicate to Client

January 2, 1970

Industrial Design Service Association Inc.
200 Middle Road
Chicago, Illinois

Attention: Mr. Ronald Smith

Gentlemen:

Confirming our understanding, IDSA agrees to perform the design/development of project B-1200, as outlined below. It's understood the project will consist of: the design and final development of a related group of three power lawn mowers, using a high degree of common parts and utilizing the basic AMCORP engine and drive train. You are to do the original design concept, creative thinking, design layouts, clay models, and finished presentation models.

Concept design sketches will be ready for presentation, February 10, 1970, and upon approval, you will immediately go into clay models, which will be ready for viewing in six weeks. Again, upon presentation to management, provide for a three dimensional study and after receiving approval, you will make the finished presentation models in approximately four weeks additional time.

You have broken down your part of the program into various phases. These, probably, should be times when we ought to get together to discuss progress, etc. Phase I—Design Concept, Phase II—Layout, Phase III—Clay Model Design, Phase IV—Detailing, and Phase V—Presentation Models.

For evaluation purposes, we will send you an engineering prototype model, to be returned after Phase I and you will purchase a Johnson Lawn King Model 40 for chassis dimensional studies. An extensive materials study will be undertaken to determine whether or not other materials, such as structural foam could be used on the economy model to advantage.

Because of the urgency of this program, you will handle finances on a project basis. As agreed, the price will be $18,000 for the three design prototypes. This will cover all costs, other than travel, and, of course, the unit you purchase, at our request, for evaluation purposes.

Any major engineering changes will, of course, alter the timing schedule, and, depending on the changes, will reopen the project cost for re-evaluation.

All rights to designs and ideas developed by IDSA, in connection with this program, will become the sole property of AMCORP and its assigns. All disclosures made by AMCORP, during the course of the project, are given and received in confidence, for the limited purpose of executing the project and shall not be discussed by anyone, unless AMCORP gives its prior consent.

Please sign and return a copy of this letter for our files.

Sincerely,

AMERICAN MANUFACTURING CORPORATION

James Jones
Vice President, Product Planning

AGREED TO BY:

IDSA

BY: _____

TITLE: _____

DATE: _____

Appendix J: Client Prepared Contract

Form: Legal Contract

Type: Client Prepared Proposal

Designer Action Required: Sign and Return One Copy to Client

PROPOSAL
FOR A
DESIGN RESEARCH PROGRAM
BY
INDUSTRIAL DESIGN SERVICE ASSOCIATES
FOR
AMERICAN MANUFACTURING CORPORATION

A. INTRODUCTION

This will confirm our discussions of November, 1969. After a careful review by the Product Planning and Design Departments of AMCORP, I am pleased to be able to tender you the following proposal for work to be done for AMCORP by IDSA during the calendar year 1970.

B. OBJECTIVES

Five projects for research, design, and development work are embraced by this proposal. The general objective of IDSA will be to study and make recommendations to AMCORP with regard to new directions and innovations in the styling and design of electronic home entertainment products designated in each of the five following areas:

1. *Systems Display.* IDSA will consult and evaluate new methods of display to replace the conventional cathode ray tube. Holography, electroluminescence, and videoceram, as well as recent AMCORP breakthroughs, will be considered. Objective is to maximize and dramatize the potentials of each system.
2. *Pacesetters.* IDSA will examine and review new engineering developments and processes, utilizing recent AMCORP innovations. A series of meetings will be scheduled between AMCORP engineering and IDSA to evaluate this design criteria.
3. *Youth.* Receivers aimed at ages younger than today's accepted levels—indestructable and super safe. Cost reductions possible through new materials and a disposable "never service; replace" philosophy.
4. *Furniture.* A renewed attempt to create "period" pieces reflecting the 1970s. Candidates could utilize new construction techniques such as structural foams, fibreglass shells, monocoque and space frames, sandwich aluminum honey comb, and other "aircraft" approaches to design.
5. *Leisure.* Speciality receiver for particular end uses may be arrived at by combinations of existing electro-mechanical conveniences. Sets might not enjoy universal (because of unnecessary features) use, but could obtain saturation impact in the education, youth, retirement, sports, and health markets.

C. COSTS

Payment for work done on the above projects, which shall be conducted during the first of the year, shall be made in the following manner. Six monthly payments of $7,000 each, commencing on January 31, 1970, and on the last day of the five succeeding months.

D. SCHEDULE

A design and development project for the second half of 1970 will be agreed upon in writing. This project will complement AMCORP'S own efforts in planning for products to be sourced off-shore and introduced in 1971.

E. GENERAL CONDITIONS

AMCORP will notify IDSA of the starting dates and completion dates for each project area. No projects are to be started prior to the initial starting date specified, unless the early starting date has received prior approval in writing from AMCORP. No changes, additions, or deletions from any of the projects as outlined generally above will be made by IDSA without first receiving prior written approval from AMCORP. It is agreed that all rights to designs and ideas developed by IDSA in connection with any of the projects covered by this agreement will become the sole property of AMCORP and its assigns. It is understood that all disclosures made by AMCORP, during the course of the several projects covered by this agreement, are given and received in confidence for the limited purpose of executing the projects there covered, and shall not be discussed by you with anyone, or disclosed to anyone, without the prior consent of AMCORP.

This agreement shall be governed by the laws of the State of New York.

F. CONCLUSION

If the foregoing meets with your approval, please sign and return the copy of this letter to us for our file.

AMERICAN MANUFACTURING CORPORATION

BY _____

James Jones
Vice President, Product Planning

APPENDIX J

Accepted and Agreed to:

INDUSTRIAL DESIGN SERVICE ASSOCIATES, INC.

BY _____

TITLE _____

DATE _____

Appendix K: Consultant-Prepared Contract

Form: Legal Contract

Type: Standard Time and Materials Agreement

Client Action Required: Sign and Return Duplicate with Purchase Order

The following standard time and materials agreement can be used by your firm to place employees of your firm with a client or to place a subcontractor or another independent consultant to work with your client. The appendix to this time and materials contract is a clause which would prevent your client from trying to hire an employee or subcontractor of yours. This is important in case you are dealing with an employee or subcontractor that one of your clients might want to hire on a full-time basis. Since this happens quite frequently, the use of this clause is quite important.

STANDARD TIME AND MATERIALS AGREEMENT

THIS AGREEMENT made as of _____, 19___, between _____("Purchaser") and STEVEN P. TOMCZAK AND ASSOCIATES ("Contractor"), 7969 W. 4th Street, Los Angeles, CA 90048.

1. *Services.* Contractor agrees to perform for Purchaser the services listed in the Scope of Services Section in Appendix "A" attached hereto. Such services are hereinafter referred to as "Services." If any provisions of Appendix "A" conflict with any provisions set forth in this Agreement, the provisions of Appendix "A" shall govern.

2. *Rate of Payment for Services.* Purchaser agrees to pay Contractor for services in accordance with the following schedule:

   Category   *Normal Week*   *Hourly Rate*

   The normal weekly rate applies to those weeks in which an individual is scheduled to work 40 hours. It is expected that more than 40 hours will be worked in some weeks but this is considered to be a characteristic of the profession.

   Extended work week rates are determined by increasing the normal weekly rate by $\frac{1}{40}$ for each hour scheduled in excess of the normal 40 hours. In weeks in which the employee works less than 40 hours, such as a holiday, the rate is reduced by $\frac{1}{40}$ for each hour less than 40 hours.

   Overtime hours are compensated in accordance with applicable law.

3. *Reimbursement for Expenses.* Contractor shall be reimbursed by Purchaser for all reasonable expenses incurred by the Contractor in the performance of Services, including jet tourist air travel, auto rental expense and per-diem living expenses of Contractor's employees while away from the Contractor's principal offices, long distance calls and computer time and supplies. Local travel from the Contractor's facility will be in accordance with standard company policy. All travel by Contractor's personnel shall be in accordance with Contractor's standard policy governing travel and business expenses.

4. *Invoicing.* Purchaser shall pay the amounts agreed to herein within _____ days of receipt of invoices submitted by Contractor. Invoices will be sent monthly/bi-monthly while Contractor is performing Services and a final invoice will be sent upon termination. Contractor shall keep sufficient records at Contractor's principal place of business to allow Purchaser to ascertain correctness of the invoices.

5. *Confidential Information.* Each party ("such party") shall hold in trust for the other party ("such other party") and shall not disclose to any non-party, any confidential information of such other party.

Confidential information is information which relates to such other party's research, development, trade secrets, or business affairs but does not include: (i) information known to such party prior to negotiations leading to this Agreement; and (ii) information which is known or able to be ascertained by a non-party of ordinary skill in the field of endeavor being considered.

6. *Contractor's Employees.* Neither Contractor nor Contractor's employees are or shall be deemed employees of the Purchaser. Contractor shall take appropriate measures to insure that its employees who perform Services are competent to do so, that they do not breach Paragraph 5 hereof and that they are adequately covered by Workmen's Compensation Insurance. Upon receipt of written notice by Contractor that an employee of Contractor is not suitable to Purchaser, Contractor shall remove such employee from the performance of Services.

7. *Rights in Work Product.* Purchaser shall have ownership of all materials and ideas embodied therein resulting from the Services and construed to be confidential information within the meaning of Paragraph 5 hereof.

8. *Termination.* Services of this agreement terminate ten (10) working days after Contractor's receipt of written notice from Purchaser that the Services shall be terminated.

    Purchaser will be liable for charges for Services including written documentation of results of Services and reimbursement for expenses until termination.

9. *Applicable Law.* Contractor shall comply with all applicable laws in performing Services but shall be held harmless for violation of any governmental procurement regulation to which it may be subject but which is not referred to in Appendix "A." This Agreement shall be construed in accordance with the laws of the State of California.

10. *Additional Work.* For ten business days after receipt of an oral order which modifies or adds to the Services, Contractor may take reasonable action and expend reasonable amounts of money based on such oral order and Contractor shall be paid for such action and expenditure on the same basis as set forth in this Agreement.

11. *Equal Employment Opportunity.* "The equal employment opportunity clause, Section 202, Executive Order 11246, as amended, related to equal employment opportunities and implementing rules and regulations of the Secretary of Labor are incorporated herein by specific reference."

12. *Notices.*
    (i) Notices to Purchaser should be sent to:
    _____
    _____

    (ii) Notices to Contractor should be sent to:
    STEVEN P. TOMCZAK AND ASSOCIATES
    7969 W. 4th Street, Los Angeles, CA 90048

APPENDIX K

IN WITNESS WHEREOF, the parties hereto have signed this Agreement as of the date first written above.

_____   _____

By _____   By _____

APPENDIX A

"Services" described herein are to be determined by management of "Purchaser" without prior specification beyond the right of "Purchaser" to interview and select employees of "Contractor" to provide such "Services."

APPENDIX B

STEVEN P. TOMCZAK AND ASSOCIATES offers a waiver of the Agreement clause which states:

> "No attempt will be made, either by the Purchaser or the Contractor, to hire any employee of either agreeing company for one (1) year following termination of said Agreement."

STEVEN P. TOMCZAK AND ASSOCIATES will activate such a waiver on an individual basis. Notification should be given prior to employment of each such individual so that proper screening as to intent and general salary requirements can be accomplished. This type of waiver is known as a twelve (12) month contract to hire.

A twelve (12) month contract to hire is based on the premise that after twelve (12) months employment under a standard T & M Agreement, the normal placement fees for direct (captive) employment have been satisfied. The Purchaser may then extend an offer of employment to the Contractor's employee at no charge to the Purchaser.

Further, each month of continuous T & M employment satisfies one-twelfth ($1/12$) of the total Contractor's placement fee, i.e.:

One (1) month T & M satisfies one-twelfth ($1/12$) of placement fee.
Two (2) months T & M satisfies two-twelfths ($2/12$) of the placement fee.
Three (3) months T & M satisfies three-twelfths ($3/12$) of placement fee, etc.

This type of Time and Material Agreement allows the Purchaser and the Contractor's employee to evaluate each other completely before a permanent arrangement is concluded.

At no time will an offer of employment be extended to any Purchaser's employee by the Contractor.

Placement Fee: Placement fees will be at the rate of one (1%) percent per one thousand dollars of annual income. Example: Employee's income equals $20,000 × 1% per thousand equals 20% equals $4,000 fee.

Appendix L: Consultant-Prepared Contract

Form: Legal Contract

Type: Agreement Between Consulting Firm and Subcontractor

Action Required: Both Consulting Firm Representative and Subcontractor Should Sign the Agreement and Fill Out Sections A, B, and C

When working for one of your clients, you might be requested to bring in extra assistance, which you can do by bringing in one of your employees. However, if you do not have an employee, you might want to retain the use of another consulting firm or independent contractor. If this is the case, you can use this standard agreement form which is an agreement between your consulting firm and an independent contractor or subcontractor who will be providing services for your client under your company name.

Note that even though the independent contractor or subcontractor is in business for himself, he is contracting to you to perform services for your client. When he performs these services, he is doing so under your company name, not under his company name. He will only be using his company name as a subcontractor to you.

Agreement Between Consulting Firm and Subcontractor

This Agreement is made effective the _____ day of _____ , 19____, between STEVEN P. TOMCZAK AND ASSOCIATES and _____(Consultant), whose Employer Identification Number/Social Security Number (cross out whichever is inapplicable) is _____.

1. *Definitions.*
    a. STEVEN P. TOMCZAK AND ASSOCIATES is a California based business with offices located at 7969 W. 4th St., Los Angeles, California 90048.
    b. "Consultant" is defined as the individual or entity which agrees to perform, or cause to be performed, the services required pursuant to this Agreement and any schedules attached hereto.
    c. "Client" is defined as any individual or entity for whom Consultant provides and performs the consulting services requested by STEVEN P. TOMCZAK AND ASSOCIATES of Consultant as described herein.
    d. "Compensation" is defined as the fee paid Consultant by STEVEN P. TOMCZAK AND ASSOCIATES for Consultant's satisfactory performance of the services described herein.
2. *Nature of Services.* Consultant agrees to perform for STEVEN P. TOMCZAK AND ASSOCIATES the consulting services described in Section A of the attached Schedule. Such services shall be performed during the period mentioned in Section B of the attached Schedule at all times and locations specified in the attached Schedule, or as defined by Client.
3. *Compensation.* STEVEN P. TOMCZAK AND ASSOCIATES shall pay to Consultant the Compensation or rate of Compensation provided in Section C of the attached Schedule. This Compensation is/is not (cross out whichever is inapplicable) renegotiable during the period of performance described in Section B of the attached Schedule. Compensation shall be paid within fifteen (15) days after STEVEN P. TOMCZAK AND ASSOCIATES' receipt and approval of Consultant's statement of services or invoice to STEVEN P. TOMCZAK AND ASSOCIATES, together with authorized and signed time cards by Client and payment by Client to STEVEN P. TOMCZAK AND ASSOCIATES for said services rendered by Consultant.
    Additional information required by STEVEN P. TOMCZAK AND ASSOCIATES (if none, so state):
    _____
    _____
    _____
4. *Independent Contractor Status.* The services to be rendered for and on behalf of STEVEN P. TOMCZAK AND ASSOCIATES to Client

shall be subject to the control of STEVEN P. TOMCZAK AND ASSOCIATES merely to the extent of the result to be accomplished by the work and not as to the means and methods for accomplishing the result.

In performing services under this Agreement, Consultant shall operate as and have the status of an independent contractor and shall not act as or be an agent or employee of STEVEN P. TOMCZAK AND ASSOCIATES.

Consultant's activities will be at his own risk and Consultant shall not be entitled to Workmen's Compensation or similar benefits or other insurance protection provided by STEVEN P. TOMCZAK AND ASSOCIATES; on the contrary, Consultant will make his own arrangements for payment of hospital and medical costs in connection with any injury or illness and other insurance coverages for the activities to be performed hereunder.

5. *Performance of Duties*. Consultant agrees to perform consulting services with that standard of care, skill and diligence normally provided by a professional person in the performance of such consulting services in respect to work similar to that hereunder. Consultant is hereby given notice that STEVEN P. TOMCZAK AND ASSOCIATES will be relying on the accuracy, competence, and completeness of Consultant's services hereunder in utilizing the results of such services in fullfilling contractual commitments to the clients of STEVEN P. TOMCZAK AND ASSOCIATES.

Consultant agrees to at all times serve and promote STEVEN P. TOMCZAK AND ASSOCIATES' interests, expectancies, and relationships with Client, and with all other customers of STEVEN P. TOMCZAK AND ASSOCIATES, and shall do nothing to disturb or devalue those interests, expectancies, or relationships, or to bring discredit upon STEVEN P. TOMCZAK AND ASSOCIATES. Without limitation on the foregoing, Consultant expressly agrees to do nothing to interfere with STEVEN P. TOMCZAK AND ASSOCIATES' rights, expectancies, or relationships with Client in connection with the work to be performed hereunder.

With STEVEN P. TOMCZAK AND ASSOCIATES' written consent, Consultant agrees, during a period of three hundred sixty-five (365) days following the termination of this Agreement, not to render services to or for the benefit of Client, directly or indirectly, whether as an employee, independent contractor, or otherwise, which services are similar to, or are substantially based on or are the consequence of, services rendered by Consultant under this Agreement.

8. *Property Rights*. STEVEN P. TOMCZAK AND ASSOCIATES shall have a permanent, assignable, nonexclusive, royalty-free license to use any concept, product, or process, patentable or otherwise, furnished or supplied to STEVEN P. TOMCZAK AND AS-

SOCIATES by Consultant, or otherwise developed by Consultant in the performance of this Agreement. If requested by STEVEN P. TOMCZAK AND ASSOCIATES, Consultant agrees to do all things necessary, at STEVEN P. TOMCZAK AND ASSOCIATES' sole cost and expense, to obtain patents or copyrights on any processes, products, or writings developed or produced by Consultant in the performance of this Agreement, to the extent that same may be patented or copyrighted, and further agrees to execute such documents as may be necessary to implement and carry out the provisions of this paragraph. All materials prepared or developed by Consultant hereunder, including documents, calculations, maps, sketches, notes, reports, data, models, and samples, shall become the property of STEVEN P. TOMCZAK AND ASSOCIATES when prepared, whether delivered to STEVEN P. TOMCZAK AND ASSOCIATES or not, and shall, together with any materials furnished Consultant by STEVEN P. TOMCZAK AND ASSOCIATES hereunder, be delivered to STEVEN P. TOMCZAK AND ASSOCIATES upon request and, in any event, upon termination of this Agreement.

7. *Nondisclosure.* Consultant agrees not to divulge to third parties, without the written consent of STEVEN P. TOMCZAK AND ASSOCIATES, any information obtained from or through STEVEN P. TOMCZAK AND ASSOCIATES in connection with the performance of this Agreement unless (a) the information is known to Consultant prior to obtaining same from STEVEN P. TOMCZAK AND ASSOCIATES, (b) the information is, at the time of disclosure by Consultant, then in the public domain, or (c) the information is obtained by Consultant from a third party who did not receive same, directly or indirectly, from STEVEN P. TOMCZAK AND ASSOCIATES and/or Client. Consultant further agrees not to disclose to any third party, without the prior written consent of STEVEN P. TOMCZAK AND ASSOCIATES, any information developed or obtained by Consultant in the performance of this Agreement, except to the extent that said information falls within one of the categories described in (a), (b), or (c) above.

8. *Nonassignment.* The services to be performed by Consultant pursuant to this Agreement are personal. Consultant shall not assign this Agreement without the prior written consent of STEVEN P. TOMCZAK AND ASSOCIATES.

9. *Waiver or Modification Ineffective Unless in Writing.* It is agreed that no waiver or modification of this Agreement or of any covenant, condition or limitation herein contained shall be valid unless in writing and duly executed by the party to be charged therewith and that no evidence of any waiver or modification shall be offered or received in evidence in any proceeding, arbitration, or litigation between the parties hereto arising out of or affecting this Agreement, or the rights or obligations of any party hereunder, unless

such waiver or modification is in writing duly executed as aforesaid, and the parties further agree that the provisions of this paragraph may not be waived except as herein set forth.

10. *Applicable Law.* This Agreement shall be governed by the laws of the State of California.

11. *Attorney's Fees and Costs.* If any action, in law or in equity is necessary to enforce or interpret the terms of this Agreement, the prevailing party shall be entitled to reasonable attorney's fees and costs in addition to any other relief to which such prevailing party may be entitled.

12. *Termination of Agreement.* STEVEN P. TOMCZAK AND ASSOCIATES reserves the right to terminate, without prior notice, this Consulting Agreement for cause, including, without limitation, STEVEN P. TOMCZAK AND ASSOCIATES' and/or Client's dissatisfaction with Consultant's services.

13. *Additional Provisions.* This Agreement also includes additional provisions, if any, as are specifically set forth in Section D and E of the attached Schedule.

STEVEN P. TOMCZAK AND ASSOCIATES
By _____
Title _____

CONSULTANT
_____
_____
Address
_____
City, State, Zip Code
_____
Telephone

SCHEDULE

Section A. Scope of Work

Section B. Period of Performance

This Agreement shall be effective as of the date first set forth herein and shall continue through _____.

Section C. Compensation

The sum of $_____ per _____ Consultant is engaged in performing the services described in Section A above.

APPENDIX L

EXTENSION OF CONSULTING AGREEMENT

The parties hereto agree to extend the terms of that certain Consulting Agreement dated _____, by and between STEVEN P. TOMCZAK AND ASSOCIATES, 7969 W. 4th Street, Los Angeles, CA 90048, and _____.

The terms of this extension shall be from _____ to _____. All terms and conditions of such Consulting Agreement shall be extended, except as follows:

Executed this ____day of _____, 1981, at _____, _____.

CONSULTANT:

_____

STEVEN P. TOMCZAK AND ASSOC.

By _____

By _____

Date _____

Appendix M:
Standard Form Agreement Between Engineer and Consultant for Professional Services

Form: Legal Contract

Type: Time and Materials

Client Action Required: Issue a Letter of Authorization or a Purchase Order

If you plan to work as a consultant on a project, you might find the following standard form agreement helpful. This standard contract is used for consultants who will only participate in a limited manner on a particular project. The National Society of Professional Engineers provides a large number of consulting contracts for many applications. The one that is included in this appendix is the contract that is used in most applications by independent contractors.

> This document has important legal consequences; consultation with an attorney is encouraged.

STANDARD FORM OF AGREEMENT
BETWEEN
ENGINEER AND CONSULTANT
FOR
PROFESSIONAL SERVICES

(Intended for use with Consultants for special services. For continuous services use Engineer-Associate Engineer Agreement, N. 1910-13, or Engineer-Architect Agreement, No. 1910-10.)

Prepared by
ENGINEERS' JOINT CONTRACT DOCUMENTS COMMITTEE
and
Issued and Published Jointly by
PROFESSIONAL ENGINEERS IN PRIVATE PRACTICE
A practice division of the
NATIONAL SOCIETY OF PROFESSIONAL ENGINEERS
and by
AMERICAN CONSULTING ENGINEERS COUNCIL
and by
AMERICAN SOCIETY OF CIVIL ENGINEERS

Copyright 1980 National Society of Professional Engineers, 2029 K Street, N.W., Washington, D.C. 20006; American Consulting Engineers Council, 1015 15th Street, N.W., Washington, D.C. 20005; American Society of Civil Engineers, 345 East 47th Street, New York, N.Y. 10017.

Guide Sheet for Completing Standard Form of
Agreement Between Engineer and Consultant
for Special Consulting Services

1. *General*—This form of agreement is intended for use for services of those consultants who will only participate to a limited extent in the design aspects of Project: such consultants might include archeologists, psychologists, value engineers, acoustical, kitchen, library and traffic consultants, landscape architects and accountants. When contracting for geotechnic services or subsurface investigations this form may be used with appropriate modifications; however, such agreements are preferably between the Owner and the geotechnical engineer. When contracting for services that are to be rendered on a continuing basis during most of the design aspects of the Project use Standard Form of Agreement between Engineer and Associate Engineer, No. 1910-13, or Standard Form of Agreement between Engineer and Architect, No. 1910-10.

2. *Page 1*—In the description of the Project insert as detailed information as possible concerning the part of the Project to which Consultant's services will pertain, such as special requirements as to performance, capacity or function, budgetary limitations and any of the special aspects or peculiarities of the Project. Identify studies, reports or analyses which Engineer has furnished to Consultant for his guidance. In the general description of Consultant's services for This Part of the Project include a general statement of the scope of his responsibilities.

3. *Exhibit A*—It is important to attach as Exhibit A all parts of the Prime Agreement that have a bearing on Consultant's services or compensation.

4. *Exhibit B*—This Exhibit is referred to in Sections 1, 2, 4, 5 and 6. It must be carefully prepared to describe in as much detail as practicable the nature of Consultant's Basic Services and the time within which such services are to be rendered. Exhibit B should itemize those Additional Services which may be requested and have been identified at the signing of the Agreement. Exhibit B must also determine the applicable methods of payment for Consultant's services. It must fix the amounts, deductible provisions and duration of Engineer's and Consultant's professional liability insurance coverages. The details of the coordination between Consultant's services and those of Engineer and his other consultants should be spelled out in detail in Exhibit B.

    A sample or suggested form of Exhibit B is attached to this Guide Sheet.

5. *Section 5—Methods of Compensation*. There are any number of ways of paying for consulting services. Section 5 contemplates that the method or methods selected will be inserted in Exhibit B in

paragraph 2 for Basic Services and in paragraph 4 for Additional Services. The most commonly used methods of payment are Lump Sum, Unit Price, Percentage of Construction Cost, Cost Plus, Hourly Rate, Per Diem Rate, Direct Labor Costs times a factor and Payroll Costs times a factor. Suggested language applicable to each of these methods for guidance in completing Exhibit B appears [after the exhibit].

6. *Paragraph 5.3—Reimbursable Expenses.* Modifications may be necessary to adapt these provisions to the particular type of consulting services involved. It may be desirable to include expenses incurred for computer time and other highly specialized equipment, for previously established programs or for photographic production techniques. Note that Exhibit B must include the number of copies of his documents that Consultant will be required to furnish without additional compensation. Note also that Section 1 requires the Consultant to obtain at his own expense all data and information (other than that referred to in paragraphs 3.1 and 3.2) necessary for the performance of his services and that paragraph 5.5.1 requires the Consultant to furnish progress reproductions and information at his own expense.

7. *Paragraph 6.4—Construction Budget.* If there is a budgetary limitation on the Construction Cost of what the Consultant is to design, the figure must be inserted here. If there is no such budgetary limitation, the paragraph should be crossed out and initialed. Note that the Construction Budget figure to be inserted in paragraph 6.4 may be adjusted from time to time and the adjusted figure confirmed in writing. The Construction Budget and the Construction Cost for This Part of the Project referred to in subparagraph 5.6.1 are not necessarily the same figure since they do not always involve the same concept.

8. *Paragraph 6.5—Insurance.* Exhibit B should indicate the amounts of the Engineer's and Consultant's professional liability insurance and the amounts of the applicable deductibles. In addition, the period of time when each is required to maintain such insurance should be indicated. If Engineer and Consultant are insured by different carriers, the extent of their respective coverages should be examined. If either party requests limitation of liability provisions, this should be separately negotiated and discussed with legal counsel as to its enforceability.

9. *Paragraph 6.8—Arbitration.* This provides for compulsory and binding arbitration of all disputes between Engineer and Consultant where the claimed amount at issue is not more than $200,000. It also restricts joinder in the arbitration proceedings of others who are not a party to the Agreement; this would, therefore, exclude the Owner and other design professionals as well as the Contractor. The Engineers' Joint Contract Documents Committee believes that arbitration of such disputes under the Construction Industry Arbi-

APPENDIX M

tration Rules of the American Arbitration Association is in the best interest of both Engineer and Consultant but also recognizes that others may differ, and the laws of all states are not similar. Accordingly, the form of Agreement has been prepared so that paragraph 6.9, Arbitration, may be eliminated from the Agreement simply by tearing out the separate page on which it appears AND completing the information at the bottom of the last page to indicate the page number as well as the total number of pages that make up the entire Agreement.

10. *Paragraph 7.2* identifies two Exhibits: Exhibit A, the Prime Agreement or pertinent portions thereof; and Exhibit B "Description of Basic Consulting Services and Related Matters". The number of pages of each Exhibit should be inserted in the blank spaces. Other Exhibits should be listed in the space provided. If Exhibit A or B is not to be attached, reference to the document should be crossed out of the Agreement.

[For further discussion of the Agreement between Engineer and Consultant, see Commentary on Contract Documents, No. 1910-9 (1980 Edition), Paragraph B1(f).]

APPENDIX M

(Illustrative Form)

EXHIBIT B TO STANDARD FORM OF AGREEMENT BETWEEN ENGINEER AND CONSULTANT FOR PROFESSIONAL SERVICES dated 1980 (for use with No. 1910-14, 1980 Edition).

Description of Basic Consulting Services and Related Matters

This is an exhibit attached to, made a part of and incorporated by reference in the Agreement made on   , 198   between
(Engineer)
and   (Owner)
providing for professional services.

1. Consultant shall provide for Engineer under Section 1 of the Agreement, the following Basic Services in accordance with the terms and conditions of the Agreement:

   Basic Services, Section 1.

   1.1 Prepare preliminary layout of recommended landscaping and attend meeting with Owner. Furnish three copies of layout.

   1.2 Prepare final layout and estimated cost for submittal to Engineering and Owner for review. Furnish three copies of layout and estimate.

   Services under paragraphs 1.1 and 1.2 will be completed within 60 days of date of the Agreement.

   1.3. Based on Owner's final approval as submitted by Engineer, prepare planting plan and specifications as requested by Engineer to be incorporated into the Project specifications. Furnish three copies of planting plan and specifications.

   Services under this paragraph 1.3 will be completed within 30 days of Engineer's request therefor.

   1.4. Upon completion of planting, visit job-site, inspect the plant material and planting work for conformance with the Consultant's design concept and compliance with the requirements of the Contractor's agreement with the Owner. Submit to Engineer in duplicate a written report with recommendations for accepting or rejecting Contractor's work and for approving payment.

   Services under this paragraph 1.4 will be completed expeditiously.

2. Compensation for Basic Services of principals and employees of Consultant rendered pursuant to Section 1 will be on the following bases:

   2.1. For services pursuant to paragraphs 1.1 and 1.2 compensation will be paid in a lump sum of $.

   2.2. For services pursuant to paragraph 1.3 compensation will be on the basis of Payroll Costs times a factor of   for services rendered by principals and employees assigned to This Part of the Project. (Include definition of Payroll Costs and identify principals.)

- **2.3.** For services pursuant to paragraph 1.4 compensation will be in an amount equal to $ for each hour of time spent on performing such services.

3. Consultant shall provide for Engineer under Section 2 of the Agreement, the following Additional Services in accordance with the terms and conditions of the Agreement:

Additional Services, Section 2.

- **3.1.** Visit nurserymen's properties to inspect plant material proposed to be delivered.
- **3.2.** Participate in opening of bids and evaluation of bidders.
- **3.3.** Be present at job-site during planting of trees and shrubs.

4. Compensation for Additional Services of principals and employees of Consultant rendered pursuant to Section 3 will be on the following bases:

- **4.1.** For services pursuant to paragraph 3.1 an amount equal to $ for each day or part thereof spent in performing such services.
- **4.2.** For services pursuant to paragraphs 3.2 and 3.3 an amount equal to $ for each hour of time spent on performing such services.

5. Engineer and Consultant shall each procure and maintain professional liability insurance in accordance with paragraph 6.4.2 of the Agreement as follows:

	Amount	**Deductible**	**Effective Through**
Engineer	$	$	
Consultant	$	$	

SUGGESTED LANGUAGE FOR DIFFERENT METHODS OF PAYMENT

Paragraph 5 of the Engineer-Consultant Agreement indicates that the Consultant's compensation for Basic and Additional Services will be as indicated in paragraphs 2 and 4, respectively, of Exhibit B to that Agreement. Suggested language applicable to each of the more customary methods of payment for consulting services is set forth below for guidance in completing paragraphs 2 and 4 of Exhibit B.

1. *Lump Sum Method of Payment:* Compensation for Basic and Additional Services of principals and employees of Consultant rendered pursuant to Sections 1 and 2 will be paid in a lump sum of $
2. *Unit Price Method of Payment:* Compensation for Basic and Addi-

tional Services of principals and employees of Consultant rendered pursuant to Sections 1 and 2 will be paid in an amount based on the following unit price schedule for services:

_____	$_____
_____	$_____
_____	$_____
_____	$_____

3. *Percentage Method of Payment:* Compensation for Basic and Additional Services of principals and employees of Consultant rendered pursuant to Sections 1 and 2 will be paid in an amount equal to _____ % of the Construction Cost of This Part of the Project.

   (Note: The term "Construction Cost for This Part of the Project" is defined in paragraph 5.6.1 of the Agreement.)

4. *Cost-Plus Method of Payment:* Compensation for Basic and Additional Services of principals and employees of Consultant rendered pursuant to Sections 1 and 2 will be paid in an amount equal to Consultant's costs directly related to the performance of his services as agreed to by Engineer and Consultant plus a fee of $ _____ for overhead and profit.

   (Note: It may be desirable to agree in advance on the costs that are and are not includable. For a detailed definition of such costs see paragraphs 11.4 and 11.5 of the Standard General Conditions of the Construction Contract, No. 1910-8 (1978 Edition).)

5. *Hourly-Rate Method of Payment:* Compensation for Basic and Additional Services of principals and employees of Consultant rendered pursuant to Sections 1 and 2 will be paid in an amount equal to $ _____ for each hour of time spent in performing such services.

   (Note: Use a tabulation when several persons or grades of personnel with different hourly rates will be involved.)

6. *Per Diem Method of Payment:* Compensation for Basic and Additional Services of principals and employees of Consultant rendered pursuant to Sections 1 and 2 will be paid in an amount equal to $ _____ for each day or part thereof spent in performing such services.

   (Note: Use a tabulation when several persons or grades of personnel with differing per diem rates will be involved.)

7. *Direct Labor Costs Times a Factor Method of Payment:* Compensation for Basic and Additional Services of principals and employees of Consultant rendered pursuant to Sections 1 and 2 will be paid in an amount based on Consultant's Direct Labor Costs times a factor of for services rendered by principals and employees engaged directly on This Part of the Project. Direct Labor Costs used as a basis for such payment mean salaries and wages (basic and incentive) paid to all personnel of Consultant engaged directly on This Part of the Project, but does not include indirect payroll-related costs or fringe

benefits. For the purposes of this Agreement the principals of Consultant and their hourly Direct Labor Costs are:

_____
_____

(Note: The individuals in Consultant's office who are considered "principals" should be identified to avoid misunderstandings later on.)

8. *Payroll Costs Times a Factor Method of Payment:* Compensation for Basic and Additional Services of principals and employees of Consultant rendered pursuant to Sections 1 and 2 will be paid in an amount based on Consultant's Payroll Costs times a factor of _____ for services rendered by principals and employees engaged directly on This Part of the Project. Payroll Costs used as a basis for such payment mean the salaries and wages (basic and incentive) paid to all personnel of Consultant engaged directly on This Part of the Project, plus the cost of customary and statutory benefits including, but not limited to, social security contributions, unemployment, excise and payroll taxes, workers' compensation, health and retirement benefits, sick leave, vacation and holiday pay applicable thereto. For purposes of this Agreement, the principals of Consultant and their hourly Payroll Costs are:

_____

The amount of customary and statutory benefits of all other personnel will be considered equal to _____ % of salaries and wages.

(Note: The individuals in Consultant's office who are considered "principals" should be identified to avoid misunderstandings later on. In lieu of the detailed accounting required to substantiate the amount paid for customary and statutory benefits of personnel, a percentage of salaries and wages may be agreed to in advance and inserted in the blank space in the last sentence; otherwise the sentence should not be used.)

CONTENTS

IDENTIFICATION OF THE PARTIES AND GENERAL
DESCRIPTION OF THE PROJECT   315

SECTION 1—BASIC SERVICES OF CONSULTANT   315

SECTION 2—ADDITIONAL SERVICES OF CONSULTANT   316

SECTION 3—ENGINEER'S RESPONSIBILITIES   316

SECTION 4—PERIOD OF SERVICE   317

SECTION 5—PAYMENTS TO CONSULTANT   317

    5.1. Method of Compensation, 317

    5.2. Future Adjustments, 318

    5.3. Reimbursable Expense, 318

    5.4. Times of Payment, 318

    5.5. Reproductions and Information, 319

    5.6. Construction Costs, 319

SECTION 6—GENERAL CONSIDERATIONS   319

    6.1. Termination, 319

    6.2. Reuse of Documents, 320

    6.3. Records, 320

    6.4. Construction Budget, 320

    6.5. Insurance, 321

    6.6. Controlling Law, 322

    6.7. Successors and Assigns, 322

    6.8. Delegation of Duties, 322

    6.9. Arbitration, 323

SECTION 7—SPECIAL PROVISIONS, EXHIBITS
AND SCHEDULES   324

    7.1. Special Provisions, 325

    7.2. Exhibits and Schedules, 325

> This document has important legal consequences; consultation with an attorney is encouraged.

STANDARD FORM OF AGREEMENT
BETWEEN
ENGINEER AND CONSULTANT
FOR
PROFESSIONAL SERVICES

THIS IS AN AGREEMENT made as of   , 198 ,
between   (ENGINEER)
and   (CONSULTANT).
ENGINEER has made an agreement dated   , 198   with
                                                 (OWNER)
which is herein referred to as the Prime Agreement and which provides for ENGINEER's furnishing professional services in connection with the Project described therein. ENGINEER hereby engages CONSULTANT to furnish for ENGINEER certain of those services in accordance with the terms and conditions of this Agreement. A copy of all portions of the Prime Agreement pertinent to CONSULTANT's responsibilities, compensation and timing of services hereunder is attached, made a part hereof and marked Exhibit A. CONSULTANT has been furnished a copy of OWNER's latest program for the Project to the extent available. The Project is described in the Prime Agreement as follows:

The part of the Project for which CONSULTANT is to furnish services is hereinafter called "This Part of the Project" and is generally described as follows:

ENGINEER is the prime professional with respect to CONSULTANT's services to be performed under this Agreement and is responsible for co-ordinating CONSULTANT's services with the services of others involved in the Project. CONSULTANT is ENGINEER's independent consultant for This Part of the Project, responsible for the means and methods used in performing consulting services under this Agreement, and is not a joint-venturer with ENGINEER.

ENGINEER and CONSULTANT agree as set forth below:

SECTION 1—BASIC SERVICES OF CONSULTANT

CONSULTANT shall provide for ENGINEER the basic consulting services described in detail in Section 1, *Basic Services* of Exhibit B "Description of Basic Consulting Services and Related Matters" within the time periods stipulated therein. Basic Services will be paid for by ENGINEER as indicated in Section 5 hereof. The CONSULTANT shall at CONSULTANT's own expense obtain all data and information (other than that referred to in paragraphs 3.1 and 3.2) necessary for the performance of his services. CONSULTANT is responsible to see that the documents prepared by CONSULTANT and the services CONSULTANT renders hereunder conform to the regulations, codes and special requirements of the place where the Project is located. All of CONSULTANT's communications to or with OWNER or ENGINEER's other consultants will be through or with the knowledge of ENGINEER.

SECTION 2—ADDITIONAL SERVICES OF CONSULTANT

If authorized in writing by ENGINEER, CONSULTANT shall furnish Additional Services which are in addition to Basic Services. To the extent that the Additional Services have been identified at the time of signing this Agreement, they are itemized in paragraph 3 of Exhibit B "Description of Basic Consulting Services and Related Matters" and will be paid for by ENGINEER as indicated in Section 5 hereof. As further Additional Services are requested by ENGINEER, this Agreement will be supplemented to describe them and indicate the method of compensation therefor.

SECTION 3—ENGINEER'S RESPONSIBILITIES

ENGINEER shall:

3.1. Provide all criteria and full information as to OWNER's requirements for This Part of the Project, including design objectives and constraints, space, capacity and performance requirements, flexibility and expandability, and any budgetary limitations; furnish copies of all design and construction standards which OWNER will require to be included in the Drawings and Specifications.

3.2. Place at CONSULTANT's disposal Drawings, Specifications, schedules and other information which were prepared by ENGINEER, or by others which is available to ENGINEER, and which ENGINEER considers pertinent to CONSULTANT's responsibilities hereunder, on all of which CONSULTANT may rely in performing services hereunder except as may be specifically noted otherwise in writing.

3.3. Request OWNER to make all provisions for CONSULTANT to enter upon public and private property as required for CONSULTANT to perform services under this Agreement.

3.4. Consult with CONSULTANT before issuing interpretations or clarifications of documents furnished by CONSULTANT, and obtain written consent of CONSULTANT before acting upon shop drawings, samples or other submittals of construction contractors or change orders affecting This Part of the Project, and assume full responsibility for any such action taken without such consultation or consent.

3.5. Prior to acceptance of any contractor or subcontractor proposed for This Part of the Project, consult with CONSULTANT to determine if CONSULTANT, after due investigation, has reasonable objection to any such contractor or subcontractor.

3.6. Furnish to CONSULTANT a copy of bidding documents and such other construction contract data as pertain to CONSULTANT's services.

3.7. Give prompt written notice to CONSULTANT whenever ENGINEER observes or otherwise becomes aware of any development that affects the scope or timing of CONSULTANT's services, or any defect in the work of contractor(s) affecting This Part of the Pronject.

3.8. Bear all costs incident to compliance with the requirements of this Section 3.

SECTION 4—PERIOD OF SERVICE

CONSULTANT recognizes that the services of ENGINEER and others involved in the Project are dependent upon the timely performance of CONSULTANT's services. CONSULTANT shall perform such services in the same character, timing and sequence as ENGINEER is required to perform services under the Prime Agreement. The dates by which the various aspects of CONSULTANT's Basic Services are to be completed are set forth in paragraph 1 of Exhibit B "Description of Basic Consulting Services and Related Matters".

SECTION 5—PAYMENTS TO CONSULTANT

5.1. *Method of Compensation.* ENGINEER shall pay CONSULTANT for Basic Services rendered under Section 1 as more particularly described in paragraph 1 of Exhibit B "Description of Basic Consuting Services and Related Matters" a fee computed as indicated in paragraph 2 of said Exhibit B; and shall pay CONSULTANT for Additional Services rendered under Section 2 as more particularly described in paragraph 3 of said Exhibit B on the basis indicated in paragraph 4 of said Exhibit B.

5.2. *Future Adjustment.* If the scope of This Part of the Project is changed materially or if the period of time during which CONSULTANT is required to render services hereunder is extended beyond   , 198   the amount of compensation provided for herein shall be adjusted appropriately.

5.3. *Reimbursable Expense.* In addition to the payments provided for in paragraph 5.1, ENGINEER shall pay CONSULTANT the actual cost of all Reimbursable Expenses incurred in connection with all Basic or Additional Services. Reimbursable Expenses mean the actual expenses incurred, directly or indirectly, in connection with This Part of the Project for: transportation and subsistence incidental thereto; obtaining bids, proposals or quotations from contractor(s); furnishing and maintaining field office facilities; toll telephone calls and telegrams; reproduction of reports, Drawings and Specifications and similar Project-related items in addition to those required by paragraph 5.5.1 and Exhibit B; and, if authorized in advance by ENGINEER, overtime work requiring higher than regular rates.

5.4. *Times of Payment.* Payments to CONSULTANT shall be made in accordance with this paragraph 5.4.

    5.4.1. CONSULTANT may submit monthly statements for Basic and Additional Services rendered and for Reimbursable Expenses incurred. When compensation is on the basis of a lump sum or a percentage of Construction Cost for This Part of the Project, the statements will be based upon CONSULTANT's estimate of the proportion of the total services actually completed at the time of billing. ENGINEER shall make prompt payments in response to CONSULTANT's statements (subject to the provisions of paragraph 5.4.2).

    5.4.2. If ENGINEER objects to any statement submitted by CONSULTANT, ENGINEER shall so advise CONSULTANT in writing giving reasons therefor within fourteen days of receipt of such bill. ENGINEER shall bill OWNER monthly on account of CONSULTANT's services and expenses and shall pay CONSULTANT within fourteen days of the time ENGINEER receives payment from OWNER on account thereof. It is intended that payments to CONSULTANT will be made as ENGINEER is paid by OWNER under the Prime Agreement and that ENGINEER shall exert reasonable and diligent efforts to collect prompt payment from OWNER. However, whether or not OWNER pays ENGINEER in full, ENGINEER shall pay all amounts due CONSULTANT within a reasonable time after completion of ENGINEER's services under the Prime Agreement.

    5.4.3. If ENGINEER fails to make any payment due CONSULTANT for services and expenses within sixty days after receipt of CONSULTANT's statement therefor, CONSULTANT may, after giving seven days' written notice to ENGINEER, suspend services under this Agreement until he has been paid in full all amounts due him for services and expenses.

5.5. *Reproductions and Information.*

    5.5.1. CONSULTANT shall at CONSULTANT's expense furnish ENGINEER five copies of all progress reproductions and information required by ENGINEER for performance of ENGINEER's services under the Prime Agreement or for review of CONSULTANT's services while in progress.

    5.5.2. ENGINEER shall at ENGINEER's expense furnish information and progress reproductions of ENGINEER's work and that of others assigned to the Project as may be required for the orderly performance of CONSULTANT's services.

5.6. *Construction Cost.*

    5.6.1. The Construction Cost for This Part of the Project means the portion of the Construction Cost of the entire Project to OWNER (determined in accordance with the Prime Agreement) which is applicable to This Part of the Project; but it does not include compensation for professional services or related expense, the cost of land, or rights-of-way, or compensation for or damages to properties unless this Agreement so specifies, nor will it include OWNER's administrative legal, accounting, insurance counseling or auditing services, or interest and financing charges incurred in connection with the Project. Labor furnished by OWNER for the Project will be included in such Construction Cost at current market rates based on CONSULTANT's estimate and shall include reasonable allowance for overhead and profit. Materials and equipment furnished by OWNER will be included at current market prices.

    5.6.2. When Construction Cost of This Part of the Project is used as a basis for payment it will be based on one of the following sources. Prior to receipt of a bona fide bid from a qualified bidder for the entire Project, or, if the work is not to be bid, prior to receipt of a bona fide negotiated proposal, such Construction Cost shall be based on CONSULTANT's most recent opinion of probable Construction Cost for This Part of the Project. After receipt of a bona fide bid or negotiated proposal for the entire Project, Construction Cost for This Part of the Project shall be determined by ENGINEER and CONSULTANT. No deduction is to be made from CONSULTANT's compensation on account of penalty, liquidated damages, or other amounts withheld from payments to contractor(s).

SECTION 6—GENERAL CONSIDERATIONS

6.1. *Termination.*

    6.1.1. The obligation to provide further services under this

Agreement may be terminated by CONSULTANT upon seven days' written notice to ENGINEER in the event of substantial failure by ENGINEER to perform in accordance with the terms hereof through no fault of CONSULTANT. It may also be terminated by ENGINEER with or without cause upon seven days' written notice to CONSULTANT. In the event of any termination, CONSULTANT will be paid for services rendered to the date of termination plus unpaid Reimbursable Expenses.

6.1.2. This Agreement will terminate automatically upon termination of the Prime Agreement. ENGINEER will promptly notify CONSULTANT of such termination.

6.2. *Reuse of Documents.* All documents furnished by CONSULTANT pursuant to this Agreement are instruments of CONSULTANT's services in respect of This Part of the Project. They are not intended or represented to be suitable for reuse by ENGINEER or others on extensions of the Project or on any other project. Any reuse without specific written verification and adaptation by CONSULTANT for the specific purposes intended will be at user's sole risk and without liability or legal exposure to CONSULTANT. Any such verification and adaptation will entitle CONSULTANT to further compensation at rates to be agreed upon by ENGINEER and CONSULTANT.

6.3. *Records.*

6.3.1. Records of CONSULTANT's Direct Labor Costs, Payroll Costs and Reimbursable Expenses pertaining to This Part of the Project will be kept on a generally recognized accounting basis and made available to ENGINEER on request.

6.3.2. CONSULTANT shall maintain all design calculations on file in legible form. A copy of these shall be available to ENGINEER at ENGINEER's expense and the originals shall not be disposed of by CONSULTANT until after sixty days' prior written notice to ENGINEER.

6.3.3. CONSULTANT's records and design calculations will be available for examination and audit if and as required by the Prime Agreement.

6.4. *Construction Budget.*

6.4.1. The portion of the total estimated Construction Cost of the Project which after discussion with CONSULTANT has been allocated by ENGINEER to This Part of the Project is $; such amount is referred to herein as the Construction Budget. The Construction Budget may be adjusted by agreement of ENGINEER and CONSULTANT from time to time on the basis of design considerations and probable Construction Cost opinions or estimates made during the course of design of the Project, and all such adjustments shall be confirmed in writing. CONSULTANT shall periodically advise ENGINEER of CONSULTANT's evaluation of the probable Construction Cost of This Part of the Project and continuously monitor CONSULTANT's work so that any facilities CONSULTANT designs for ENGINEER hereunder will be within the Construction Budget.

6.4.2. Estimates and statements of probable construction costs prepared by CONSULTANT are to represent CONSULTANT's best judgment based on experience and qualifications. It is recognized, however, that CONSULTANT does not have any control over the cost of labor, materials or equipment, over contractors' methods of determining prices, or over competitive bidding or market conditions. Accordingly, CONSULTANT cannot and does not guarantee that proposals, bids or the actual Construction Cost for This Part of the Project will not vary from any Statement of Probable Construction Cost or other cost estimate prepared by CONSULTANT. Any Construction Budget established by ENGINEER will include a bidding contingency of ten percent unless another amount is agreed upon in writing; and CONSULTANT after consultation with ENGINEER shall be permitted to determine what materials, equipment, component systems and types of construction are to be included in the Contract Documents with respect to This Part of the Project, and to make reasonable adjustments in the scope of This Part of the Project to bring it within the Construction Budget. If required, CONSULTANT shall assist ENGINEER in including in the Contract Documents provisions for alternative bids to adjust the Construction Cost to the Construction Budget.

6.5. *Insurance.*

6.5.1. ENGINEER and CONSULTANT shall each procure and maintain insurance for protection from claims under workers' compensation acts, claims for damages because of bodily injury including personal injury, sickness or disease or death of any and all employees or of any person other than such employees, and from claims or damages because of injury to or destruction of property including loss of use resulting therefrom.

6.5.2. Also ENGINEER and CONSULTANT shall each procure and maintain professional liability insurance for protection from claims arising out of performance of professional services caused by any negligent error, omission or act for which the insured will provide for coverage in such amounts, with such deductible provisions and for such periods of time as set forth in paragraph 5 of Exhibit B "Description of Basic Consulting Services and Related Matters"; and certificates indicating that such insurance is in effect will be exchanged by them.

 6.5.3. ENGINEER will also cause other professional consultants retained by ENGINEER for the Project to procure and maintain comparable professional liability insurance coverage.

6.6. *Controlling Law.* This Agreement is to be governed by the law of the principal place of business of ENGINEER.

6.7. *Successors and Assigns.*

 6.7.1. ENGINEER and CONSULTANT each is hereby bound, and the partners, successors, executors, administrators, assigns and legal representatives of each are bound, to the other party to this Agreement and to the partners, successors, executors, administrators, assigns and legal representatives of such other party, in respect to all covenants, agreements and obligations of this Agreement.

 6.7.2. Neither ENGINEER nor CONSULTANT shall assign, sublet or transfer any rights under or interest in (including, but without limitation, moneys that may become due or moneys that are due) this Agreement without the written consent of the other, except as stated in paragraph 6.7.1 and except to the extent that the effect of this limitation may be restricted by law. Unless specifically stated to the contrary in any written consent to an assignment, no assignment will release or discharge the assignor from any duty or responsibility under this Agreement.

 6.7.3. Nothing herein shall be construed to give any rights or benefits hereunder to anyone other than ENGINEER and CONSULTANT.

6.8. *Delegation of Duties.* If in this Agreement it is stated that the Basic Services of CONSULTANT are to be performed by one or more specified individuals within CONSULTANT's organization, only the individuals so specified shall perform services hereunder and their duties shall not be delegated to any other individual or entity without the written consent of ENGINEER. If not so stated, CONSULTANT may employ such other consultants, associates and subcontractors as CONSULTANT may deem appropriate for assistance in the performance of services hereunder.

[The remainder of this page was left blank intentionally.]

6.9. *Arbitration.*

6.9.1. All claims, counterclaims, disputes and other matters in question between the parties hereto arising out of or relating to this Agreement or the breach thereof will be decided by arbitration in accordance with the Construction Industry Arbitration Rules of the American Arbitration Association then obtaining, subject to the limitations and restrictions stated in paragraphs 6.9.3 and 6.9.4 below. This agreement so to arbitrate and any other agreement or consent to arbitrate entered into in accordance herewith as provided in this paragraph 6.9 will be specifically enforceable under the prevailing arbitration law of any court having jurisdiction.

6.9.2. Notice of demand for arbitration must be filed in writing with the other parties to this Agreement and with the American Arbitration Association. The demand must be made within a reasonable time after the claim, dispute or other matter in question has arisen. In no event may the demand for arbitration be made after institution of legal or equitable proceedings based on such claim, dispute or other matter in question would be barred by the applicable statute of limitations.

6.9.3. All demands for arbitration and all answering statements thereto which include any monetary claim must contain a statement that the total sum or value in controversy as alleged by the party making such demand or answering statement is not more than $200,000 (exclusive of interest and costs). The arbitrators will not have jurisdiction, power or authority to consider, or make findings (except in denial of their own jurisdiction) concerning any claim, counterclaim, dispute or other matter in question where the amount in controversy thereof is more than $200,000 (exclusive of interest and costs) or to render a monetary award in response thereto against any party which totals more than $200,000 (exclusive of interest and costs).

6.9.4. No arbitration arising out of, or relating to, this Agreement may include, by consolidation, joinder or in any other manner, any person or entity who is not a party to this Agreement.

6.9.5. By written consent signed by all the parties to this Agreement and containing a specific reference hereto, the limitations and restrictions contained in paragraphs 6.9.3 and 6.9.4 may be waived in whole or in part as to any claim, counterclaim, dispute or other matter specifically described in such consent. No consent to arbitration in respect of a specifically described claim, counterclaim, dispute or other matter in question will constitute consent to arbitrate any other claim, counterclaim, dispute or other matter in question which is not specifically described in such consent or in which the sum or value in controversy exceeds $200,000 (exclusive of interest and costs) or which is with any party not specifically described therein.

6.9.6. The award rendered by the arbitrators will be final, not subject to appeal and judgment may be entered upon it in any court having jurisdiction thereof; and will not be subject to modification or appeal except to the extent permitted by Sections 10 and 11 of the Federal Arbitration Act (9 U.S.C. §§ 10, 11).

[The remainder of this page was left blank intentionally.]

APPENDIX M

SECTION 7—SPECIAL PROVISIONS, EXHIBITS AND SCHEDULES

7.1. *Special Provisions.* This Agreement is subject to the following special provisions.

    7.1.1.

7.2. *Exhibits and Schedules.* The following Exhibits are attached to and made a part of this Agreement:

    7.2.1. Exhibit A—Copy of portions of Prime Agreement consisting of   pages.

    7.2.2. Exhibit B "Description of Basic Consulting Services and Related Matters" consisting of   pages.

    7.2.3.

7.3. This Agreement (consisting of pages 1 to   , inclusive) together with the Exhibits identified above constitute the entire agreement between ENGINEER and CONSULTANT and supersede all prior written or oral understandings. This Agreement and said Exhibits and schedules may only be amended, supplemented, modified or cancelled by a duly executed written instrument.

IN WITNESS WHEREOF, the parties hereto have made and executed this Agreement as of the day and year first above written.

ENGINEER   CONSULTANT

_____   _____

_____   _____

_____   _____

Appendix N: Client-Prepared Contract

Form: Consulting Services Agreement Legal Contract
Type: Time and Materials
Client Action Required: None

In many cases you will find yourself being hired by a large engineering or data processing company that will have you sign a standard consulting services agreement. These agreements are very similar from company to company. Therefore, I have included a sample of one from Hughes Aircraft Company to give you an idea of the type of clauses that are included in this type of consulting services agreement.

Note that with some companies that deal with the Government and the Department of Defense there is a clause on Security Clearance requirement. Prior to taking on a contract with a large firm you should always review the consulting services agreement to make sure that there are no clauses in it that would limit you in the future. Consulting service agreements such as the one included in this appendix are usually combined with a purchase order from the company that describes what rates you will be charging, the time limit of the contract, and so forth.

HUGHES

HUGHES AIRCRAFT COMPANY

PURCHASE ORDER ATTACHMENT NUMBER TSC-2

CONSULTANT SERVICES CONTRACT

I   PERFORMANCE: Seller shall use its best efforts; shall perform the work in accordance with the highest standards; and shall effect completion of each assigned task on or before the date specified, if any.

II PAYMENT: Unless different intervals are stipulated in this Agreement, payment for the services performed by Seller shall be made monthly, at the rates prescribed, upon submission by the Seller of properly certified invoices, including time statements, on or before the 10th day of the month next following the month in which such services were performed. In addition to the foregoing, Seller shall be reimbursed for reasonable travel expenses incurred at the request of Buyer. Such request must be submitted in writing to Seller prior to the incurrence of any such expense and signed by the Company's representative charged with the direct responsibility for Seller's performance hereunder. Authorized travel expenses are as stipulated in this Agreement.

III TIME: Time shall be of the essence hereunder; however, Seller shall not be liable for any delay in performance due to causes beyond Seller's reasonable control and without Seller's fault or negligence.

IV SECURITY: When access to Buyer's facility is required by Seller in the performance of services under this Agreement, Seller must secure and execute the Buyer's necessary security forms furnished by the Buyer for facility access. These forms must be returned at least twenty-four (24) hours prior to the desired access. If access to classified information is required in performance of the services hereunder, Seller shall meet the security clearance requirements of the U. S. Government as set forth in the current edition of the Industrial Security Manual for Safeguarding Classified Information (Attachment to DD Form 441). In the event such security clearance requirements are not met by Seller, this Agreement shall be of no force or effect. Seller will assure that classified information received, generated or reproduced during the performance period will be safeguarded at all times in accordance with the provisions of the current edition of the DoD security manual (Attachment to DD Form 441).

6-76 Edition

PURCHASE ORDER ATTACHMENT NUMBER TSC-2

V ACCESS TO PLANT PROPERTY: Without limiting Seller's obligations under Clause IV above, Seller shall comply with all the rules and regulations established by Buyer for access to and activities in and around Buyer's plants and properties.

VI RELATIONSHIP: Seller will, at all times during the performance of this Agreement and in connection with any services rendered by Seller to Buyer, be considered an independent contractor. No relationship of employer-employee is created by this Agreement or by Seller's service. Seller hereby acknowledges that Buyer is not obligated to provide Workmen's Compensation Insurance covering Seller's personnel or any other employee insurance or benefits of Buyer. Seller is notified that Buyer considers the Federal Insurance Contributions Act and the withholding provisions of the Federal or State Revenue Codes as not being applicable to any payments by Buyer to Seller pursuant to this Agreement.

VII INVENTIONS AND INFORMATION:

A. Seller shall promptly disclose to Buyer in writing, all inventions, developments, improvements and discoveries (whether or not patentable) which Seller, or any of its personnel, makes or conceives, either solely or jointly with others, which relate to the performance of this Agreement or which result from knowledge of Buyer's activities obtained by virtue of performance hereunder; excepting, however, inventions (patented or unpatented) which Seller has made or conceived and has disclosed in writing to others prior to commencing performance of this Agreement.

B. At all times during the performance of this Agreement and thereafter, whenever requested so to do by Buyer, Seller shall execute and deliver to Buyer any and all applications, assignments, and other instruments which may be necessary in order to apply for and obtain or protect, for Buyer's benefit, letters patent of the United States and foreign countries covering said inventions, developments, improvements or discoveries and which may be necessary to assign and convey to Buyer or its nominee the sole and exclusive right, title and interest therein. These obligations shall be binding upon Seller's assigns, executors, administrators or other legal representatives.

6-76 Edition

PURCHASE ORDER ATTACHMENT NUMBER TSC-2

    C. Seller shall not, except as necessary in the performance of this Agreement or as authorized in writing by Buyer, supply, disclose, or otherwise permit access to, or authorize any other person to supply, disclose, or otherwise permit access to at any time any information concerning or in any way related to inventions, information or other matter pertaining to Buyer's business which Seller may in any way acquire by reason of performance of this Agreement.

    D. Seller agrees to obtain appropriate binding agreements with any person performing work on behalf of Seller relating to the performance of this Agreement, specifying that they shall assume towards it and Buyer all of the obligations and responsibilities that Seller by Clauses VII, VIII, and IX assumes towards Buyer.

VIII COPYRIGHTS:

    A. Seller shall grant to Buyer and to its officers, agents and employees acting within the scope of their employment:

        1. A royalty-free, non-exclusive and irrevocable license to reproduce, deliver, perform, translate, publish, use, and dispose of, and to authorize others so to do, all copyrightable material first produced or composed and delivered to Buyer under this Agreement by Seller; and

        2. A license as aforesaid under any and all copyrighted or copyrightable work not first produced or composed by the Seller in the performance of this Agreement but which is incorporated in the material furnished under this Agreement; provided that such license shall be only to the extent Seller now has, or prior to completion or final settlement of this Agreement may acquire, the right to grant such license without becoming liable to pay compensation to others solely because of such grant.

    B. Seller shall exert all reasonable effort to advise Buyer, at the time of delivering any copyrightable or copyrighted work furnished under this Agreement, of any adversely held copyrights or copyrightable material incorporated in any such work and of any invasion of the right of privacy therein contained.

6-76 Edition

HUGHES
HUGHES AIRCRAFT COMPANY

PURCHASE ORDER ATTACHMENT NUMBER TSC-2

    C. Seller shall report to Buyer, promptly and in reasonable written detail, any notice or claim of copyright infringement received by Seller with respect to any material delivered under this Agreement.

IX INFORMATION/REPORTS: It is mutually understood and agreed that any and all information developed or supplied or reports submitted hereunder are solely for Buyer and shall not be divulged by Seller to other parties verbally or in writing or reproduced without prior written approval of Buyer. Notwithstanding any other provisions of this Agreement, Buyer retains the unilateral and unrestricted right to use the herein produced material or information in any and all ways Buyer may deem necessary.

X TERMINATIONS:

    A. In addition to any other rights of Buyer hereunder and not in limitation thereof, this Agreement may be terminated by three (3) days notice in writing, at no cost; provided, however, that all rights of the parties arising by calls or work assignment of the Buyer, under the Agreement, issued prior to the time of such termination shall survive such termination (see X.B, below).

    B. Calls or work assignments under this Agreement may be terminated by Buyer by giving written notice thereof to Seller. Upon receipt of Notice of Termination, Seller shall terminate all work and deliver to Buyer the results of Seller's performance to that time, including, without limitation, the documents called for under Clauses VII and VIII above. In the event of such termination, Buyer shall make to Seller, and Seller shall accept as full compensation hereunder, payment at the rates prescribed in this Agreement for travel expense as stipulated in this Agreement and work performed up to the time of receipt of the Notice of Termination, less any and all previous payments made, and Buyer shall thereupon be released from further obligation to make payments under this Agreement.

XI CONTROLS:

    A. The Seller shall maintain careful records in such a manner as to be able to determine at all times the exact current balance of the not-to-exceed funds in this Agreement. If at any time Seller has reason to believe that the applicable payments which will accrue in the performance of this order in the next succeeding thirty (30) days, when added to all other payments and costs previously accrued, will exceed seventy-five percent (75%) of the not-to-exceed price then set forth, Seller shall notify Buyer to that effect.

6-76 Edition

HUGHES

PURCHASE ORDER ATTACHMENT NUMBER TSC-2

    B.   Buyer shall not be obligated to pay Seller any amount in excess of the not-to-exceed price set forth in this Agreement, and Seller shall not be obligated to continue performance if to do so would exceed the price, unless and until Buyer shall have notified Seller in writing that such not-to-exceed price has been increased and shall have specified in such notice a revised not-to-exceed amount which shall thereupon constitute the price for performance of this Agreement. When and to the extent that the not-to-exceed price set forth has been increased, any time expended and authorized costs incurred by Seller in excess of the not-to-exceed price prior to the increase shall be allowable to the same extent as if such time expended and costs had been incurred after such increase.

6-76 Edition

Index

Acceptances, of offers, 109
Accountants:
   types, 155, 156
   typical rates, 156
Advertising:
   agencies, 37
   classified section of trade magazines, 38
   cooperative, 30
   technical trade magazines, 36
Agencies, consulting, 59–65
American Arbitration Association, 128, 129
Annuities, 161
Arbitration, 120, 128–130, 306
   clauses in contracts, 120, 129, 323
Attorney, retaining, 116
Audience feedback in presentations, 77
Audio visual aids, 83
Automobile expenses, 172

Bilateral contracts, 121
Breach of contract, 138
   remedies for, 139, 140
Brochures, 25–28, 212
Business licenses, 203
Business publication rates and data, 31
Business writing, improving, 205
Buy-sell agreements, 205, 206

Certificate of Insurance, 65
Client-consultant relationship, 2, 6
Close, 73
Closed corporation, 192
Communication skills, improving, 204
Compensation:
   cost plus fixed fees, 91
   cost times factor, 312
   fixed fees, 91
   lump sum, 310
   methods of, 6
   payroll cost times factor, 312
   percentage of construction costs, 91
   per diem, 91
   salary times multiplier, 91
   unit price method, 310
   using retainer, 91
Conferences, and taxes, 169
Consideration, 118–121
Consulting:
   agencies, 59–65
   agreement, extension of, 299
   definition, 3
   improving consultant/client relationship, 116
   similarities with present work, 5
   societies, 223, 225
   technical, 1
Contractor:
   independent, 3
   prime, 123
Contracts:
   agreement between consulting firm and contractors, 293
   arbitration, 323
   arbitration clause, 306
   bilateral, 121
   breach of, 138, 139
   copyright clauses, 333
   cost plus fixed fee, 153
   definitive aspects of, 117
   express, 121
   extension of consulting agreement, 299
   fixed monthly charge, 173
   general guidelines for, 231

handling incomplete performance, 134
implied, 121
insurance, 321
insurance clause, 306
invention clause, 332
misrepresentation and, 140
out of state laws and, 125
professional licensing and, 126
quasi, 122
reformation, 133
royalty agreements, 267
sample, *see Appendices*
six month charge, 261
standard time and materials agreements, 187
summary of past, 22
termination, 135
   clause, 334
   types of, 135–138
time and materials, 301, 329
unilateral, 121
valid, 117
Convention, foreign, 170
Copyright clauses in contracts, 333
Corporation, 190
   advantages of, 192
   closed, 192
   creation, 193
   double taxation, 192
   limited liability in, 191
   powers, 191
   Sub-chapter S, 194
   transfer of ownership, 192
Cost plus fixed fee, 6, 95, 96, 153
Counter offers, 109
Court, staying out of, 127
Credit, establishing, 215

DBA, 203
Deadlines, handling of contractual, 130
Deductions:
   deciding what to deduct, 164
   professional tax, 166
Delaware corporations, 195
Direct labor costs, 311
Disability insurance, 202

Employment, temporary *vs.* indefinite, 165
Employment Identification Number, 65
Engineering societies, 224, 225
Entertainment expenses, 168
Equipment leasing, 175
Experience required for consulting, 12
Express contracts, 121

False statements, 140
Federal Identification Number, 204

Fees:
   breakdown, 100–102
   consulting, 7
   determination, 87–102
   determination for specialists, 89
   salary times a multiplier, 97
   schedules, 98, 99
   see also Compensation
Fictitious name statements, 203
Financial advisor, selection, 153
Fixed fees, 6, 93–95
Fixed percentage of cost of project, 6
Fraud, 141
   remedies to, 142

Hourly rate method, 311

Implied contracts, 121
Incomplete performance, 134
Incorporation:
   articles of, 65
   costs, 193
   see also Corporation
Independent contractor, IRS definition of, 163
Informal consulting associations, 181, 182
Insurance:
   in contracts, 321
   disability, 202
   group, 201
   liability, 181, 201, 225
   life, 216
   worker's compensation, 202
Interviews, 72
   assessing client's business ethics, 70
   assessing client's viability, 70
   close, 73
   discussing background in, 70
   first few minutes, 68
   obtaining client's objectives, 69
   telephone, 68
Inventions:
   clause in contracts, 332
   taxes, 169
Invoicing, 150
IRS, Employment Identification Number, 65

Job shops, 59–65
Joint ventures, 185

Keogh Plan, 158

Lawyer, retention, 116
Letters of recommendation, 19, 21, 150
Licenses, 202
Lump sum, 310

INDEX

Mailing companies, 55
Mailing lists:
   commercial, 50
   creation of, 52–55
   questions for list companies, 52
   sources, 50, 51
Misrepresentation, 140
Mistakes:
   clerical, 133
   mutual, in technical matters, 133
   unilateral, 132

Negotiating fee, 104–109
Negotiation, 103–113
News releases, 31

Offers, handling, 109
Office in home, deductions for, 167
Overhead, 102

Parole Evidence, 112–113
Partnerships:
   advantages and disadvantages, 183
   agreements, write of, 187
   buy, sell agreements for, 216
   and consultant unpopularity, 183
   definition, 182
   general, 183
   limited, 184, 185
   termination, 189
Patents, 223
Percentage of construction costs, 96
Per diem, 6, 91, 311
Permits, 202
Personal communication, improvement, 204
Phone answering services, 211
Practice, cost of starting, 8, 9
Presentations:
   audio visual aids, 83
   effective, 75–85, 204
   sales, 81
      scheduling and costing, 82
      use of overhead projector, 84
Prime contractor, 123
Product releases, 31
Professional, licensing and contracts, 126
Professional engineer, 218
Professional Engineers Licensing Guide, 218
Promotion:
   while attending seminars, 40
   through field representatives, 31
   through teaching, 31
   at technical conferences, 41
   through technical trade magazines, 33

Publications, list of, 23
Publishing, 32, 212

Quasi contracts, 122

Rates, consulting, 87–102
Referrals, 29
Reports, technical, 23
Reputation, need for, 13
Resume, 17–20, 61, 62, 212
Retainer, 6, 30
Retainer contracts, 1, 237, 243, 249
Retirement plans, 158
Royalty agreement, 167

Sales presentations, 81
Scale, 89
Secretary:
   dictation skills, 208
   filing, 210
   office equipment, 209
   selection, 205
Sellers permits, 203
Seminars, 42
Sole proprietorship:
   limited life of, 178
   shotgun suits and, 180
   unlimited liability of, 179
Standard time and materials agreements, 287
Stationery, 212
Stockholders, 191
Straight time, 6
Sub-chapter S corporation, 194
Subcontractor, 123
   agreement between consulting firm and, 293

Taxes, 8
   automobile expenses, 172
   equipment leasing, 175
   keeping records, 165
   self-employment, 162
   seminars and, 169
   spouse's expenses, 169
Technical information pack, 24
Technical journal articles, 32
Technical write-in, 205
Temporary *vs.* indefinite employment, 165
Travel expenses, 168

Unit price, 310

Worker's Compensation, 65, 201

Zoning laws, 203